BLASTING OPERATIONS

Gary B. Hemphill, P.E.

Construction Manager
Hecla Mining Company

McGraw-Hill Book Company

New York St. Louis San Francisco Auckland
Bogotá Singapore Johannesburg London
Madrid Mexico Montreal New Delhi
Panama São Paulo Hamburg
Sydney Tokyo Paris
Toronto

Library of Congress Cataloging in Publication Data

Hemphill, Gary B
  Blasting operations.

  Includes index.
  1. Blasting. I. Title.
TA748.H45 624.1'52 80-10369
ISBN 0-07-028093-2

1234567890 KPKP 8987654321

The editors for this book were Jeremy Robinson and Geraldine Fahey, the designer was Mark E. Safran, and the production supervisor was Paul Malchow. It was set in Melior by Bi-Comp, Incorporated.

Printed and bound by The Kingsport Press.

To Patty

CONTENTS

13. Estimating 189

14. Safety 213

15. Specifications and Inspection 233

PREFACE

The purpose of this book is to familiarize engineers and contractors exposed to blasting with the basic fundamentals. This text is not a handbook or a "do-it-yourself" manual; rather, it is a discussion of basic blasting theory and those factors that affect blasting. The author's ambition is to upgrade blasting in some way by helping those people—engineers, contractors, and others—who are involved with blasting but know little about it to better understand the conditions and limitations under which an explosives engineer must perform.

No one can learn to blast by reading a book. As with flying an airplane, it is not enough to understand all the theory and safety rules; without proper instruction in the cockpit, or field, it is an extremely hazardous and foolish undertaking.

The text begins with an explanation of various types of military and commercial explosives, which will acquaint the reader with the various explosives available, their purpose, and their basic attributes (Chapter 2). Then there is a discussion of drills, compressors, and drilling techniques in Chapters 3 and 4 to help the engineer better understand the equipment requirements. The discussion of blasting theory is covered in several chapters, including those dealing with firing systems (Chapter 5), breakage theory (Chapter 6), blast design (Chapter 7), special blasting problems (Chapter 11), vibration (Chapter 10), and geology (Chapter 9). A discussion of these subjects is necessary to acquaint the reader with all the elements that affect blasting. The book also discusses the laws that regulate the industry and the recommended safety "dos and don'ts." Chapter 13, on estimating, discusses the considerations the engineer must realize when making an estimate of blasting costs.

This book is not all-inclusive. The blasting industry has developed so in the last decade that it would require volumes of material written by various specialists to cover all the variables that affect blasting projects. However, it is the goal of this author to give the reader a new understanding of the difficulties involved when using explosives in excavation. Building demolition is not discussed in any detail, because it is a specialized field beyond construction blasting and this author does not feel qualified to write on the subject with any authority.

The formulas, techniques, and opinions expressed in this book are examples of how, under certain conditions, explosives may be used. These formulas should not be taken literally or be utilized by those who are inexperienced. Blasting calculations must be done *only* by experienced blasters or explosives engineers. With this in mind, the engineer can use this book to help to develop strategies and methods of bringing projects involving blasting to successful conclusions.

Gary B. Hemphill

ONE

INTRODUCTION

NEED FOR KNOWLEDGE

Blasting and explosives touch the life of every man, woman, and child in the modern world. Every one of us consumes or uses things that somewhere in their processing required the use of explosives. Perhaps it is the use of electricity generated by the burning of coal, which is mined with the help of explosives. Or steel, which has a multitude of uses in today's environment and requires blasting to mine its ore. The highways on which we drive not only required the obvious blasting of rock cuts, but they also used an aggregate in the pavement that had to be blasted at the quarry. The list is endless.

As human beings continue to increase their claims on the land, converting once-rural areas into stylish suburbs, the demand on the blasting industry is increased. Quarries that were once concerned only with the costs of blasting are now concerned with the vibrations of blasting and possible damage to the ever-nearer suburban dwelling. Years earlier, the quarry was out in the country, and no one cared how much vibration was caused by the blasting because no one was even aware of it. However, this is all changed. Some quarry operations now face lawsuits, injunctions, and protesters at the gate. These things had little to do with producing aggregate a short time ago. Also, land that was once considered undesirable because it was a "rock pile" is now prime land because of its close proximity to town. Being close to everything is fine for property values, but it does create additional problems with blasting.

This increase in the complexity of blasting has created the need for a greater understanding by both contractors and engineers of the factors affecting blasting.

HISTORY OF DEVELOPMENT

No one is certain who first used explosives. The development of black powder has been attributed to both the Chinese and the Arabs. It is thought that the primary use for black powder was fireworks. It is not until the thirteenth century that historical records on black powder appear: the writings of Roger Bacon give instructions on how to make black powder.

Black powder was used for warfare before it was first used in 1627 in the Royal Mines of Schemnitz in Hungary. By 1689, it was being used in the tin mines at Cornwall. Black powder received limited use until the invention of the safety fuse in 1831 by William Bickford.

Alfred Nobel (1833–1896) introduced nitroglycerin dynamite in 1867. He accomplished this by absorbing nitroglycerin into collodion cotton so that it was much safer to handle than in its pure form. In 1875, Nobel invented the first gelatin by making a mixture of 92% nitroglycerin and 8% collodion cotton. This "blasting gelatine" still has limited applications today.

As we shall see in Chapter 2, explosives have been continually developed and improved upon to further decrease their inherent risks.

HOW TO USE THIS BOOK

If the engineer or construction manager has a project that requires blasting, this book may be used to help in the preparation of specifications, cost estimates, and safety requirements. One may take a blasting problem and by following this text, chapter by chapter, bring the problem to a credible solution. The engineer may develop a tentative blast design (for planning purposes), making it possible to determine the amount of drilling required, the drilling equipment needed, the crew sizes, and the amount of explosives needed, all of which have to be known to develop a reasonable cost estimate and schedule.

STATE OF THE ART

Blasting is still in its scientific development stage; that is, although it is now being studied and quantified by scientists, blasting is still very much an art. As the variables are further quantified, blasting will cease to be an art and will become a mixture of data and judgment. This further development as a science will enable the practitioner, or blaster, to do the job with more skill and certainty by helping to establish values for variables that were once left to empirical judgments.

Although this text can be used as a primer for blasting apprentices, it is intended for engineers who write specifications and design projects that will require rock excavation, mining engineers, and those engineers or contractors responsible for completing projects that use blasting.

TWO

EXPLOSIVES

CHEMICAL EXPLOSIVES

A chemical explosive is a compound or mixture which is capable of undergoing extremely rapid decomposition, thereby releasing substantial amounts of heat and gas. An explosive creates energy by releasing hot gases, which require a space many times the original volume of the explosive and therefore exert great pressure on the surroundings.

An explosion can be broken down into four phases: (1) release of gas, (2) intense heat, (3) extreme pressure, and (4) the explosion. An understanding of these phases can benefit in understanding the theory of breakage for blasting rock.

When an explosive is detonated, gas is released, and as the temperature of the gas increases, the pressure also increases (according to Charles' law). Therefore, when an explosive is detonated in a borehole, the pressure exerted against the walls of the borehole (if great enough) will move and break the rock.

Table 2.1 is a listing of the various ingredients that may be found in explosives.

CLASSIFICATION OF EXPLOSIVES

There are two general classifications of explosives: (1) commercial and (2) military. Commercial explosives are used in commerce and industry for the purpose of doing work. The most common types of industrial explosives are dynamite, water gels, and blasting agents. Military explosives are those used for warfare, primarily as propellants. An explosive is restricted to one classification or the other.

Military Explosives

It must be realized that military explosives differ quite markedly from commercial explosives in their use and therefore in their characteristics. Military explosives are more brisant and less sensitive than commercial explosives. They have to be less sensitive because they are used under such adverse conditions. For this reason nitroglycerin is not used in military explosives.

TABLE 2.1 Ingredients Used in Explosives

Ingredient	Chemical formula	Function
Ethylene glycol dinitrate	$C_2H_4(NO_3)_2$	Explosive base; lowers freezing point
Nitrocellulose (guncotton)	$C_6H_7(NO_3)_3O_2$	Explosive base; gelatinizing agent
Nitroglycerin	$C_3H_5(NO_3)_3$	Explosive base
Nitrostarch		Explosive base; "nonheadache" explosives
Trinitrotoluene (TNT)	$C_7H_5N_3O_6$	Explosive base
Metallic powder	Al	Fuel sensitizer; used in high-density slurries
Black powder	$NaNO_3 + C + S$	Explosive base; deflagrates
Pentaerythritol tetranitrate (PETN)	$C_5H_8N_4O_{12}$	Explosive base; caps, detonating cords
Lead azide	$Pb(N_3)_2$	Explosive used in blasting caps
Mercury fulminate	$Hg(ONC)_2$	Explosive used in blasting caps
Ammonium nitrate	NH_4NO_3	Explosive base; oxygen carrier
Liquid oxygen	O_2	Oxygen carrier
Sodium nitrate	$NaNO_3$	Oxygen carrier; reduces freezing point
Potassium nitrate	KNO_3	Oxygen carrier
Ground coal	C	Combustible, or fuel
Charcoal	C	Combustible, or fuel
Paraffin	C_nH_{2n+2}	Combustible, or fuel
Sulfur	S	Combustible, or fuel
Fuel oil	$(CH_3)_2(CH_2)_n$	Combustible, or fuel
Wood pulp	$(C_6H_{10}O_5)_n$	Combustible; absorbent
Lampblack	C	Combustible
Kieselguhr	SiO_2	Absorbent; prevents caking
Chalk	$CaCO_3$	Antacid
Calcium carbonate	$CaCO_3$	Antacid
Zinc oxide	ZnO	Antacid
Sodium chloride	NaCl	Flame depressant (permissible explosives)

SOURCE: C. E. Gregory, *Explosives for North American Engineers*, Trans Tech Publications, Clausthal, West Germany, 1973, p. 40.

The most common military explosive is trinitrotoluene (TNT), which has a detonating velocity of 22,600 feet per second, or ft/s (6900 meters per second, or m/s), and can be used for demolition charges, bursting charges, and booster charges. All other explosives are measured or compared against TNT. (See Table 2.2.)

Pentaerythritol tetranitrate (PETN) is a sensitive, powerful explosive that has a detonating velocity of 27,200 ft/s (8300 m/s). It is used in blasting caps and detonating cord. PETN is also used underwater, because it is insoluble in water.

Cyclonite (RDX) is a brisant and highly sensitive explosive. RDX is the primary base charge in the composition explosives used in blasting caps and has a detonating velocity of 27,400 ft/s (8350 m/s).

Tetryl has a detonation velocity of 23,300 ft/s (7100 m/s) and is used

TABLE 2.2 Characteristics of Principal U.S. Explosives Used for Demolition Purposes

Name	Principal uses	Velocity of detonation		Effec-tiveness*	Fumes	Water resistance
		m/s	ft/s			
Ammonium nitrate	Demolition charges and composition explosives	2,700	8,900	—	Dangerous	None
PETN	Detonating cords, blasting caps, and demolition charges	8,300	27,200	1.66	Slight	Excellent
RDX	Blasting caps and composition explosives	8,350	27,400	1.60	Dangerous	Excellent
TNT	Demolition charges and composition explosives	6,900	22,600	1.00	Dangerous	Excellent
Tetryl	Booster charges and composition explosives	7,100	23,300	1.25	Dangerous	Excellent
Nitroglycerin	Commercial dynamites	7,700	25,200	1.50	Dangerous	Good
Black powder	Time blasting fuses	400	1,300	0.55	Dangerous	Poor
Amatol 80/20	Bursting charges	4,900	16,000	1.17	Dangerous	Very poor
Composition A3	Booster charges and bursting charges	8,100	26,500	—	Dangerous	Good
Composition B	Bursting charges	7,800	25,600	1.35	Dangerous	Excellent
Composition C3	Demolition charges	7,625	25,000	1.34	Dangerous	Good
Composition C4	Demolition charges	8,040	26,400	1.34	Slight	Excellent
Tetrytol 75/25	Demolition charges	7,000	23,000	1.20	Dangerous	Excellent
Pentolite 50/50	Booster charges	7,450	24,400	—	Dangerous	Excellent

* Relative effectiveness as a breaching charge with TNT = 1.00.

SOURCE: "Explosives and Demolitions," U.S. Department of the Army Field Manual FM 5-25, 1971, p. 1-2.

TABLE 2.3 Characteristics of Dynamites

Name		Principal uses	Velocity of detonation		Relative effectiveness as a breaching charge	Intensity of poisonous fumes	Water resistance
			m/s	ft/s			
Military dynamite, M1		Demolition charge	6,100	20,000	0.92	Dangerous	Good
Straight dynamite (commercial)	40%	Demolition charges	4,600	15,000	0.65	Dangerous	Good (if fired within 24 h)
	50%		5,500	18,000	0.79		
	60%		5,800	19,000	0.83		
Ammonia dynamite (commercial)	40%		2,700	8,900	0.41	Dangerous	Poor
	50%		3,400	11,000	0.46		
	60%		3,700	12,000	0.53		
Gelatin dynamite (commercial)	40%		2,400	7,900	0.42	Slight	Good
	50%		2,700	8,900	0.47		
	60%		4,900	16,000	0.76		
Ammonia gelatin dynamite (commercial)	40%		4,900	16,000	—	Slight	Excellent
	60%		5,700	18,700	—		

SOURCE: "Explosives and Demolitions," U.S. Department of the Army Field Manual FM 5-25, 1971, p. 1-11.

primarily as a boosting charge. At one time tetryl was used in composition explosives, but it is gradually being replaced by PETN and RDX.

Ammonium nitrate is used in some composites because it is the least sensitive military explosive, but because of its water solubility it is not widely used. Ammonium nitrate is used primarily as a commercial product and is considered a blasting agent rather than an explosive.

Composites Composition A3 is a composite consisting of 91% RDX and 9% wax. The wax acts as a coating that both desensitizes and bonds the particles of RDX. A3, used as a booster for shape charges and bangalore torpedoes, has a detonating velocity of 26,500 ft/s (8100 m/s) and is used as the main charge in high-explosive plastic (HEP) projectiles.

Composition B is 60% RDX, 39% TNT, and 1% wax. It is more sensitive than TNT, has a great shattering effect, and with its detonating velocity of 25,600 ft/s (7800 m/s) is used in shape charges.

Newer bangalore torpedoes have as their base charge composition B4, which is a composite of 60% RDX, 39.5% TNT, and 0.5% calcium silicate.

Composition C2 is 80% RDX and 20% plasticizer. The plasticizer contains TNT and other explosive ingredients and gives C2 plastic characteristics.

C3 replaces C2 and consists of 79% RDX and 21% plasticizer. The plasticizer makes C3 maintain plastic characteristics from −20 to +125°F (−28 to +51°C). C3 is brisant and insoluble in water and has a detonating velocity of 25,000 ft/s (7625 m/s). It is used primarily for demolition charges.

C4 is a composite of 91% RDX and 9% nonexplosive plasticizers. It is malleable and pliable from −70 to +170°F (−56 to +76°C) and is more stable and less subject to erosion underwater than C2 and C3. C4 has a detonating velocity of 26,400 ft/s (8040 m/s); during the war in Vietnam it received quite a bit of unauthorized use: it was burned to heat rations.

Tetrytol is a "demo" charge that is more powerful and brisant than TNT and less sensitive than tetryl. It is a composite of 75% tetryl and 25% TNT and has a detonating velocity of 23,000 ft/s (7000 m/s).

Pentolite contains 50% PETN and 50% TNT. It has a detonating velocity of 24,400 ft/s (7450 m/s) and is used as a booster charge in some shape charges.

Military dynamite is a composite that contains 75% RDX, 15% TNT, and 10% desensitizers and plasticizers. It is equivalent in performance to 60% commercial dynamite and contains no nitroglycerin. It has primarily the same applications as its commercial counterpart, but because commercial dynamite contains nitroglycerin it is not used in combat areas. (See Table 2.3.)

TYPES OF EXPLOSIVES

There are three common types of industrial explosives used in construction today: (1) high explosives, (2) blasting agents, and (3) water gels.

High Explosives

High explosives detonate at a velocity range of 5000–25,000 ft/s (1525–7620 m/s). These explosives are generally in cartridges made of paper or plastic. High explosives are capable of undergoing a chemical reaction that converts them to gases at very high temperatures under great pressure.

High explosives used in industry have two primary bases: nitroglycerin and ammonium nitrate. Additives are used with these bases to resist freezing, to make the explosive less sensitive to shock, and, in the case of permissibles, to reduce the heat and the duration of the explosion.

Comparing Explosives

Before proceeding with any discussion of types of explosives it is necessary to clarify the methods or standards by which explosives are compared.

Strength Explosives, dynamites in particular, will generally have a rating such as 40 or 60%. The origin of this rating stems from straight-nitroglycerin

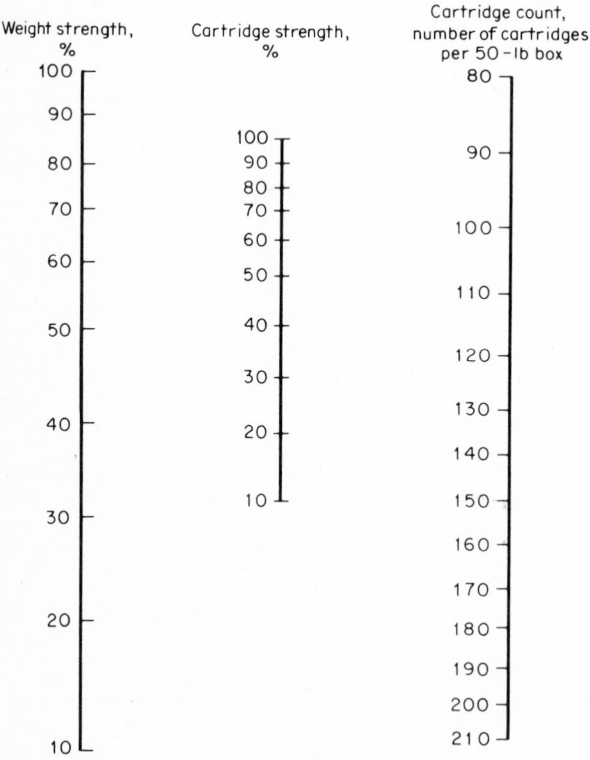

Fig. 2.1 Nomograph for comparing weight strength and cartridge strength. (*From A. B. Cummins and I. A. Given, "SME Mining Engineering Handbook," American Institute of Mining, Metallurgical, and Petroleum Engineers, Inc., New York, 1973.*)

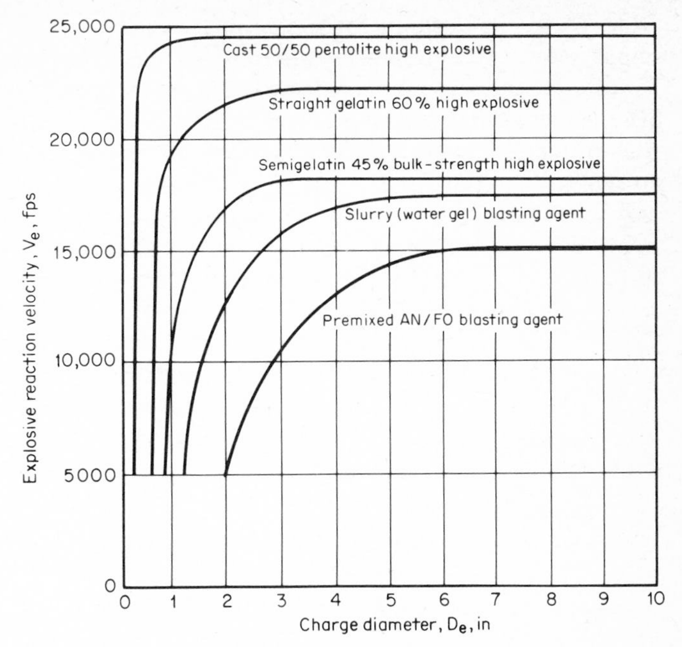

Fig. 2.2 The effects of charge diameter on detonation velocity. *(From A. B. Cummins and I. A. Given, "SME Mining Engineering Handbook," American Institute of Mining, Metallurgical, and Petroleum Engineers, Inc., New York,* 1973.)

(straight-NG) dynamite: it refers to the percent by weight of nitroglycerin in straight-NG dynamite.

This rating can be somewhat deceptive, and one should be careful not to interpret it too literally; i.e., 40% NG dynamite is not twice as strong as 20%. Since nitroglycerin is not the only energy-producing substance in the explosive compound, if the amount of nitroglycerin is increased, other energy-contributing substances must be reduced. (See Figure 2.1.)

In non–straight-NG explosives the percentage rating is a comparison to an equal weight of NG dynamite. That is, 40% ammonia dynamite can produce the same energy as 40% straight-NG dynamite.

Detonating Velocity Velocity of detonation (VOD) is the velocity at which the detonation wave moves through the explosive. It is expressed in feet per second or meters per second; and the higher the VOD, the greater the shattering effect. (See Figure 2.2.)

Fume Class The amount of toxic fumes given off by an explosive when detonated determines whether or not it is safe to use the explosive in a confined area, such as underground. The toxic gases may include carbon monoxide and oxides of nitrogen.

There are two sets of fume standards for explosives. The U.S. Bureau of Mines rates only permissible explosives, i.e., those explosives the Bureau permits in underground coal mines. The Institute of Makers of Explosives (IME) rates all nonpermissible explosives.

Permissible explosives have additives, such as sodium chloride, to produce a flame of the lowest heat, least volume, and shortest duration to reduce the possibility of secondary explosions from methane and coal dust. They have detonating velocities from 5000 to 18,000 ft/s (1525 to 5500 m/s), depending on whether the explosive is granular or gelatin in form, the granular having the slower VOD. All permissible explosives are ammonia dynamites, and they have poor storage capabilities, because they tend to absorb moisture, reducing the effectiveness of the explosive.

Commercial examples of permissible explosives are:

Manufacturer	Product name
Atlas	Coalite
CIL	Monobel
Hercules	Red HA

The Bureau of Mines classifications are class A, which is 0 to 1.87 cubic feet, or ft^3 (53 liters, or L), of noxious gases per 1½ pounds, or lb (680 grams, or g), of explosive; and class B, which is 1.87 to 3.74 ft^3 (53 to 106 L) of noxious gases per 1½ lb (680 g) of explosive. The IME classifications are class 1, for 0 to 0.16 ft^3 (0 to 0.0045 m^3) of noxious gas per cartridge; class 2, for 0.16 to 0.33 ft^3 (0.0045 to 0.0093 m^3) of noxious gas per cartridge; and class 3, for 0.33 to 0.67 ft^3 (0.0093 to 0.0189 m^3) of noxious gas per cartridge.

Water Resistance Water resistance is just what the term implies: it is the ability of explosives to resist contamination or a reduction in strength when exposed to water. Although there is not a numerical rating applied to water resistance, an explosive's water resistance is sometimes determined or demonstrated by the length of time it can be submerged in water and still perform as designed.

Density The density of an explosive is the weight of the explosive per given volume. In regard to cartridge explosives the density is generally expressed in cartridges per case. Thus, if one explosive has a cartridge count of 110 cartridges, or sticks, per 50-lb case and another has a cartridge count of 90 sticks per 50-lb case, the latter is the denser of the two.

This variance in the density of different explosives can aid in blast design, because when a great amount of energy is needed, as with tunnel rounds, an explosive with higher density permits greater energy in the same area. However, when there is no need for a greater amount of energy in a small space, the explosive energy may be spread out in the borehole by using a less dense explosive. (See Figure 2.3.)

Physical Characteristics Commercial explosives can take three basic forms: (1) granular, (2) gelatin, and (3) slurry. The form depends on the formula, and the choice of form depends on the usage required. The package for the same explosive product may also vary according to usage. For example, a slurry can be pumped into a borehole with no container, or it can be packaged in polyethylene bags to permit handling in smaller amounts.

Storage The "storability" of an explosive product means how well the product can be stored without affecting its safety, reliability, and perfor-

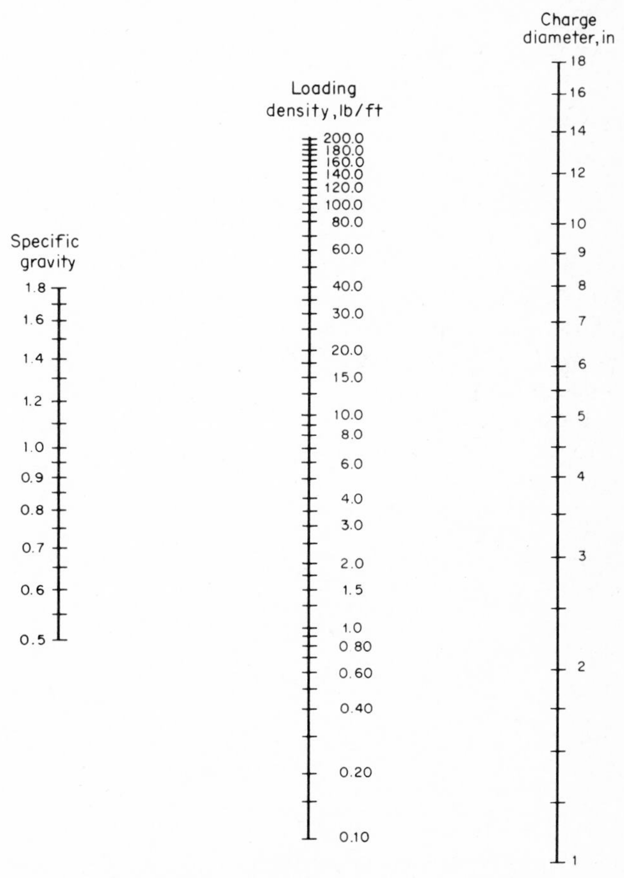

Fig. 2.3 Nomograph for finding loading density. (*From A. B. Cummins and I. A. Given, "SME Mining Engineering Handbook," American Institute of Mining, Metallurgical, and Petroleum Engineers, Inc., New York,* 1973.)

mance. Early nitroglycerin dynamites were extremely poor for storing. After a period of time, the nitroglycerin would separate from the other components and create an extremely hazardous condition.

Although the storage characteristics of explosives have greatly improved, it is still important to turn over supplies to prevent prolonged storage.

Freezing An explosive's resistance to freezing can be important for both safety and performance. Before any explosive is purchased for cold climates, the resistance to freezing and the effects of the cold on the product should be understood.

NG explosives, although they contain antifreezing additives, can become dangerous if frozen, and cap-sensitive water gels can fail to be cap-sensitive in typical winter temperatures (0 to 40°F, −17 to +4°C, depending on the product).

Types of Commercial High Explosives

The major types of commercial high explosives are (1) straight-nitroglycerin dynamite, (2) ammonia dynamites, (3) gelatin dynamites, and (4) semigelatins.

Straight-Nitroglycerin Dynamite Straight-nitroglycerin dynamite is very much like the original dynamite developed by Alfred Nobel, except that dynamites of this type now use ingredients that take part in the chemical reaction whereas Nobel mixed the nitroglycerin with kieselguhr (SiO_2, or diatomaceous earth), which added no energy. The ingredients in straight-nitroglycerin dynamite, as the name implies, are nitroglycerin and carbonaceous materials such as sawdust and nitrocellulose (guncotton). It is granular and has a high detonating velocity (17,000 ft/s, or 5200 m/s). Today, it is rarely used because of its sensitivity; however, it is well-suited for ditching and mudcapping.[1]

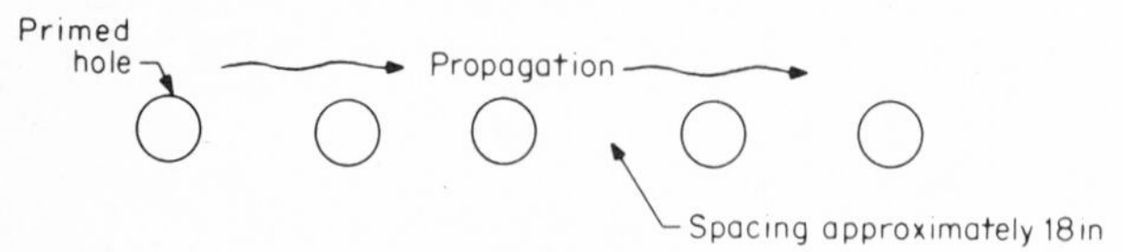

Fig. 2.4 Ditching with straight-NG dynamite by propagation.

Ditching dynamite is used to blast mostly loamy soils such as peat. The dynamite is so sensitive to shock that it can be spaced throughout the peat and the shock from detonating one hole will cause the explosion to propagate through the remaining holes. (See Figure 2.4.)

[1] For mudcapping details see Chapter 11.

Straight-nitroglycerin dynamite has good storage attributes but is in fume class 3 and should not be used for underground blasting.

Some commercial examples of straight-NG dynamite are:

Manufacturer	Product name
CIL	Ditching dynamite
Hercules	50% nitroglycerin dynamite
Independent	Straight dynamite

Ammonia Dynamites Ammonia dynamites were developed to replace straight dynamites. They are granular, and they use ammonium nitrate as a substitute for some of the liquid nitroglycerin. The ammonium nitrate is less sensitive to heat and shock, and therefore it is safer to use. It has a lower density and is also less resistant to water. Because of the low water resistance of ammonia dynamites, it is best when loading not to break the cartridge and permit moisture to enter. They are generally not recommended for underground use because of a class 3 fume classification. However, one should check individual commercial products, because fume class can vary.

Ammonia dynamites come in both high and low density, the low density being less expensive because less NG and more ammonium nitrate (AN) are used.

The detonating-velocity range for ammonia dynamite is approximately 5000–11,000 ft/s (1525–3350 m/s).

The following are some of the commercial names under which ammonia dynamites are produced:

Manufacturer	Product name
Atlas	Ammodyte (low-density)
	Extra 60% (high-density)
Hercules	Honcol (low-density)
	Extra dynamite (high-density)

Gelatin Dynamites Gelatin dynamites have a base of water-resistant "gel" made by dissolving nitrocotton in nitroglycerin. The nitrocotton gel is insoluble in water and tends to bind together other ingredients, making them water-resistant and forming a cohesive, plastic-like substance having a confined detonating velocity of 13,000 ft/s (4000 m/s).

Only the higher-strength gels (75 and 80%) are recommended for mudcapping operations. Gelatins may detonate at low velocities or fail entirely under high hydrostatic loads.

Straight gelatins (without ammonium nitrate) have good water resistance and storage characteristics. The lower strengths (20%) have detonating velocities of 14,000 ft/s (4250 m/s) and are generally of fume class 1. However, the higher-strength (90%) straight gels have a VOD of 23,000 ft/s (7000 m/s) and should not be used underground.

Ammonia gelatins are to straight gelatins as ammonia dynamites are to straight-NG dynamites. Ammonia gelatins are lower in cost but also less resistant to water. They are in fume class 1 but do not possess good storage capabilities. The most common strength used is 40%; however, you will find high-strength (75%) being used for mudcapping.

Common commercial product names are:

Manufacturer	Product name
Atlas	Giant gelatins (ammonia)
Hercules	Hercules 40% gelatin (straight)

Semigelatins Semigelatins are designed to combine the resistance to water and cohesiveness of gelatins with the lower cost of ammonia dynamites. Semigelatins tend to have poor storage and temperature-resistance characteristics. They are generally in fume class 1 and work well as presplit explosives.

Some commercial names are:

Manufacturer	Product name
Atlas	Gelodyn
CIL	Dygel
Hercules	Gelamite

See Figure 2.5.

Presplit Explosives Presplit explosives are generally ⅝ to ⅞ in in diameter and are designed to be detonated before the production shot to control overbreak (see Chapter 8). Usually a semigelatin, a presplit explosive comes either in a rigid column (⅝ in × 24 in) with couplers to enable the blaster to

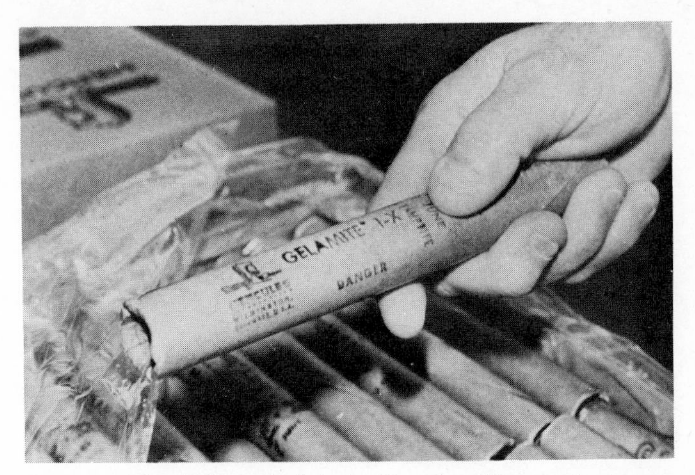

Fig. 2.5 Gelamite®, semigelatin by Hercules. (*Hercules Inc.*)

couple the tubes of explosives into a continuous column, or spaced on a line (generally detonating cord) like sausages.

Examples of commercial names of presplit explosive charges are:

Manufacturer	Product name
Apache	Smoothite
Atlas	Kleen-Kut
Austin	Red-E-Split B
Independent	Split-ex

Blasting Agents A blasting agent is a mixture containing both fuel-producing and oxidizing elements. None of the elements in a blasting agent is considered an explosive when not mixed, and blasting agents cannot be detonated by a No. 8 blasting cap when unconfined.

The most common blasting agent is nitrocarbonitrate (NCN), which is more generally known as AN/FO. AN/FO is a mixture of ammonium nitrate and fuel oil in which ammonium nitrate acts as the oxidizer and fuel oil acts as the fuel. The fuel may be something other than fuel oil, such as carbon, sawdust, hydrocarbons, or other carbonaceous substances. However, the most commonly used is ammonium nitrate with fuel oil in 94 : 6 weight ratio (or approximately 6%). The chemical reaction of AN/FO is:

$$\underset{\text{Ammonium nitrate}}{3NH_4NO_3} + \underset{\text{Fuel oil}}{CH_2} \rightarrow CO_2 + 3N_2 + 7H_2O$$

The four most common elements of a chemical explosive are hydrogen (H), carbon (C), nitrogen (N), and oxygen (O). Of these elements hydrogen and carbon are considered the fuel elements and oxygen the oxidizer. Nitrogen acts as the bonding agent that attaches itself to the oxygen and bonds the oxygen to the elements of the molecule. Explosive formulas are designed to contain an efficient balance of fuel and oxygen. If there is too little oxygen the explosive will be deficient in the amount of energy released, and there may be an undesirable by-product such as carbon monoxide (CO), which is poisonous. If there is too much oxygen the explosive will be inefficient and will not release all its potential energy and may release wasteful or dangerous oxides of nitrogen.

AN/FO offers great economy and safety in modern blasting applications. It generally costs one-quarter to one-half as much as nitroglycerin explosives and it is considerably safer to handle because of its lack of sensitivity. In many types of blasting situations AN/FO will produce better fragmentation due to its high gas-producing properties. Generally speaking, AN/FO is the best type of explosive for blasting dry boreholes in excess of 2½ in in diameter, which are conducive to breakage by gaseous expansion. However, AN/FO is poor in small-diameter boreholes and conditions that require very high detonating velocities. In Table 2.4, the detonating velocity of AN/FO can be found relative to the borehole diameter.

The primary disadvantage of AN/FO is its lack of water resistance. Once it comes in contact with water AN/FO is no longer dependable and, in fact, will not detonate.

Mixing can be a problem with AN/FO; that is, sometimes the AN and the fuel oil are not properly mixed. However, the problem can be overcome by careful mixing; or on small, non–bulk-load jobs, AN/FO may be purchased in premixed bags. (See Figures 2.6 and 2.7.)

TABLE 2.4 Confined Detonation Velocity and Borehole Loading Density of AN/FO

Borehole diameter, in	Confined VOD ft/s	Loading density, lb per foot of borehole
1.5	7,000–9,000	0.6– 0.7
2	8,500–9,900	1.1– 1.3
3	10,000–10,800	2.5– 3.0
4	11,000–11,800	4.4– 5.2
5	11,500–12,500	6.9– 8.2
6	12,000–12,800	13.3–15.8
8	12,500–13,300	17.6–20.8
9	12,800–13,500	20.0–26.8

SOURCE: C. E. Gregory, "Explosives for North American Engineers," 2d ed., Trans Tech Publications, Clausthal, West Germany, 1973, p. 59.

Fig. 2.6 Bulk loading AN/FO from bulk truck. (*Atlas Powder Co.*)

Fig. 2.7 Loading AN/FO from premixed 50-lb bags. (*Atlas Powder Co.*)

Fig. 2.8 Loading slurry in plastic cartridge. (*Atlas Powder Co.*)

Binary Explosives Binary, or two-component, explosives are not considered explosives until they are combined at the job site. Until mixed, they have no explosive properties, and thus they are not required to be shipped and stored as explosives. Before mixing, they are perfectly safe for handling and are thus easy to ship and store.

These two-component explosives consist of a powder component and a liquid component. Once they are mixed the liquid will penetrate the powder within 15 minutes (min), and if sealed in a container, the explosive has an indefinite shelf life. However, once it is completely exposed to the air, the liquid will evaporate (in approximately 15 min), leaving the powder insensitive to blasting caps.

Binary explosives have detonating velocities ranging from approximately 14,000 to 26,000 ft/s (4,200 to 8,000 m/s).

Examples of commercial names of binary explosives are:

Manufacturer	Product name
Kinetics International Corp.	Kinepar Rigid Sticks
Atlas	Kinestik
Xplo Corp.	Marine Pac Liquid

Slurries, Water Gels, Emulsions AN/FO is soluble in water and therefore, unless packaged, cannot be used under wet conditions. However, to combat this problem explosives manufacturers have developed water gels to protect the ammonium nitrate against contamination from water. The slurries that were developed to solve the AN/FO-and-water problem are commonly a mixture of an ammonium nitrate base in an aqueous solution with a combustible fuel, a heat-producing metal, and other ingredients to give a thick, soupy slurry. The slurry is 5 to 40% water and contains a gelling agent that solidifies the slurry in the borehole to protect the ammonium nitrate from water. Because of a density greater than that of water (1.05 to 1.8 grams per cubic centimeter, or g/cm) the slurry will sink to the bottom of a wet borehole. The detonation velocity of slurries ranges from 11,000 to 18,000 ft/s (3350 to 5500 m/s), which can be greater than that for AN/FO. (Refer again to Figure 2.2, a graphical comparison of the detonating velocities of various explosives relative to charge diameter.)

Recent developments in the production of slurries, or water gels, now permit the packaging of the slurries in light plastic cartridges to allow transportation and storage of slurries. (See Figure 2.8.) Also, researchers have developed cap-sensitive water gels by adding a metallic sensitizer, which permits cap sensitivity down to 20°F (−6°C). However, these cap-sensitive water gels cannot be classified as blasting agents: their behavior is similar to that of high explosives. (See Figure 2.9.)

Because of this substantial advancement in explosive products, some researchers are calling the gelative dynamites obsolete. Many blasters and most major explosive manufacturers disagree with this observation, because there are many varying grades or types of blasting conditions and rock. Each condition and rock type requires a different combination and amount of

Fig. 2.9 APEX 463 emulsion by Atlas. (*Atlas Powder Co.*)

explosives. Varying properties of the different explosives permit a proper matching of a particular explosive to a particular blasting condition. Cap-sensitive water gels will perform as well as gelative dynamites in most cases. The greatest advantage of a water gel is the lack of nitroglycerin, which adds stability and removes the cause of painful headaches: nitroglycerin, when absorbed into the bloodstream through the pores, can cause violent headaches and nausea. Many blasters feel that the best feature of cap-sensitive water gels is the lack of "powder" headaches.

Some examples of commercial water gel products are:

Manufacturer	Product name
Atlas	Aquagel (sensitized)
	RXL463 Emulsion (not sensitized)
Du Pont	Tovex
Hercules	Gel Power A (sensitized)
	Gel Power O (not sensitized)

See Figures 2.10 and 2.11.

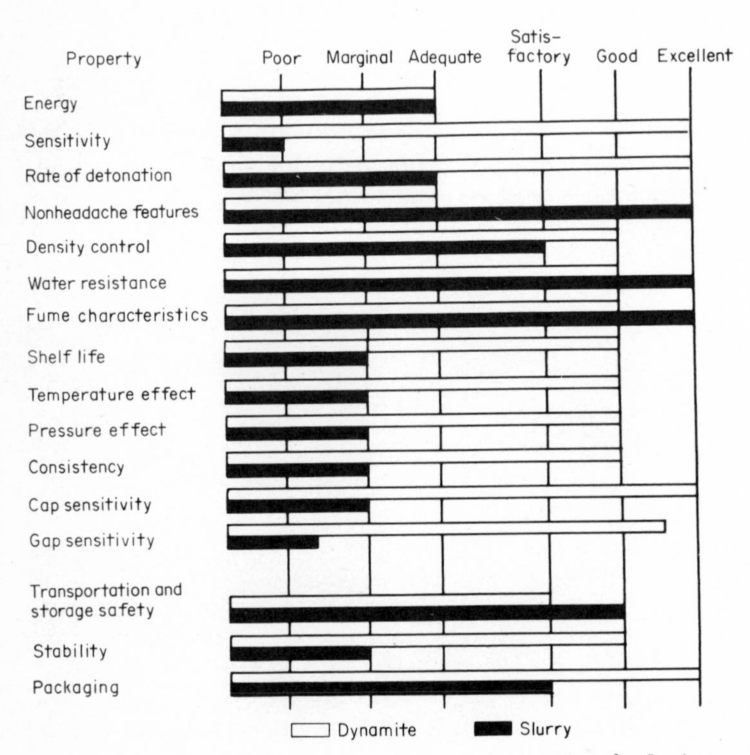

Fig. 2.10 Comparison between dynamite and slurry. (*Hercules Inc.*)

Fig. 2.11 Advertisement promoting the continued use of dynamite. (*Atlas Powder Co.*)

CONCLUSIONS

The reader has probably observed that the development of explosives from NG dynamite to sensitized water gels has been meant to improve the product and make the forerunner obsolete. For example, ammonia dynamite was supposed to replace straight-NG dynamite. However, you can still purchase straight-NG dynamite. As explosives evolve, the new and generally improved explosive takes an appreciable portion of the market of its predecessor. However, thus far, no explosive has been completely replaced. Such is the case with cap-sensitive water gels.

Cap-sensitive water gels are a great step in the evolution of commercial explosive products, and this author has used them and would certainly recommend their use. However, they should not be used exclusively as if they were the only explosive available. The blaster has a tremendous responsibility, and once the shot has been detonated the blaster has no control over the blast. Therefore, the blaster must make great efforts in the planning of the blast and use the best judgment to accomplish the task. Part of this is the selection of explosives. The blaster must match the blasting requirements and blasting methods and materials. Therefore it is important that the blaster have more than one type of explosive to choose from.

THREE

DRILLS AND DRILLING

The first process involved in a drilling and blasting operation is the drilling. The principle of drilling is to obtain maximum penetration to enable the placement of the explosive charges. There are four major types of rock drills: hand-held, external-percussion, down-the-hole percussion, and rotary.

The drilling system consists of the drill; the drill steel, or rod; and the bit. The bit penetrates the rock by the force it imposes on the rock. Bits are designed for percussion, rotary drilling, or both. The coordination of these forces with the slope, or geometric properties, of the bit is what enables the rock to be penetrated. The rotation of the bit against the bottom of the borehole creates shear stresses in the rock, causing its separation. The percussion, or hammering, chips the rock by the compressive forces under the bit. (See Figure 3.1.)

HAND-HELD DRILLS

The hand-held drill is a hand-held tool for drilling small-diameter holes (1 to 2 in, or 25 to 50 millimeters, or mm) in rock. These drills are generally limited to shallow or small amounts of drilling. Because of the time involved in drilling a hole with a hand drill, the drill is used only when the use of a larger drill is impractical. These drills generally weigh between 30 and 90 lb (13 and 40 kilograms, or kg) and are powered by compressed air. (See Figure 3.2.)

EXTERNAL-PERCUSSION DRILLS

Percussion drills combine rotation and percussion for penetrating rock. These drills (drifters) are generally mounted on a crawler undercarriage or on a truck body for construction uses. Generally the main power source for these drills is compressed air; however, the controls are generally hydraulic. The external rotary-percussion drill is generally efficient up to a depth of 60

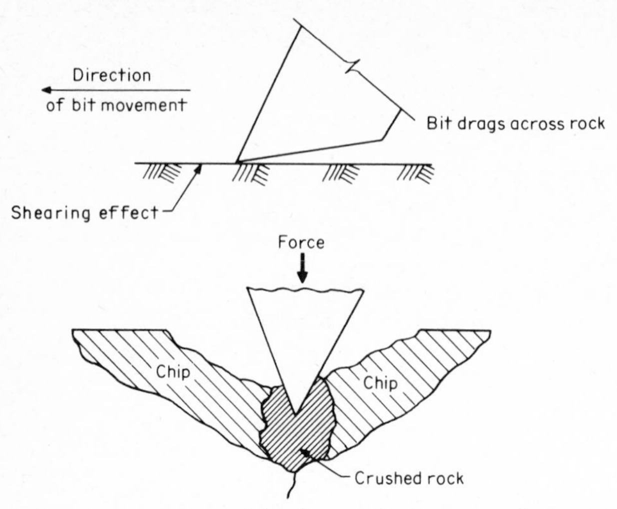

Fig. 3.1 Penetrating rock by shear and compressive forces.

ft (18 m). The external-percussion drill has a drifter containing a piston, which when subject to sufficient air pressure (usually 90 to 100 lb/in^2) will start moving reciprocally within the cylinder. On the downward stroke the piston hits the striking bar (or shank) 1200 to 2500 times per minute, and this in turn transports the energy through the drill steel to the bit penetrating the rock. (See Figure 3.3.)

The rotation is created by the use of a splined rifle bar with pawls and ratchet or by a motor controlled independently of the movement of the piston. The rifle bar, which causes the rotation of the attached drill steel as it moves vertically through the rifle nut, comes in various ratios, such as one in thirty. This ratio refers to the pitch of the flutes or splines on the rifle bar, and it indicates the length of rifle bar required for one rotation. For example, a one in forty ratio means that the rifle bar has to travel 40 in to make one complete rotation. This ratio permits the calculation of the theoretical rotational speed in revolutions per minute of the rifle bar. To illustrate, a one in fifty pitch striking bar receiving 2100 blows a minute with a 2 ½-in piston stroke will develop $(2100 \times 2.5)/50 = 105$, or approximately 100 r/min. The higher the pitch, the fewer the revolutions per minute; thus, a one in thirty pitch creates a higher rotational speed than a one in fifty pitch.

Independent Rotation

Independent rotation is just as the name implies; i.e., the rotation is independent of the reciprocatory piston, so that the rotation can be regulated without the influence of the hammer. Therefore, in soft-material drilling, where greater rotation is needed relative to percussion, the rotation may be increased and the hammer, or percussion, in blows per minute, may be de-

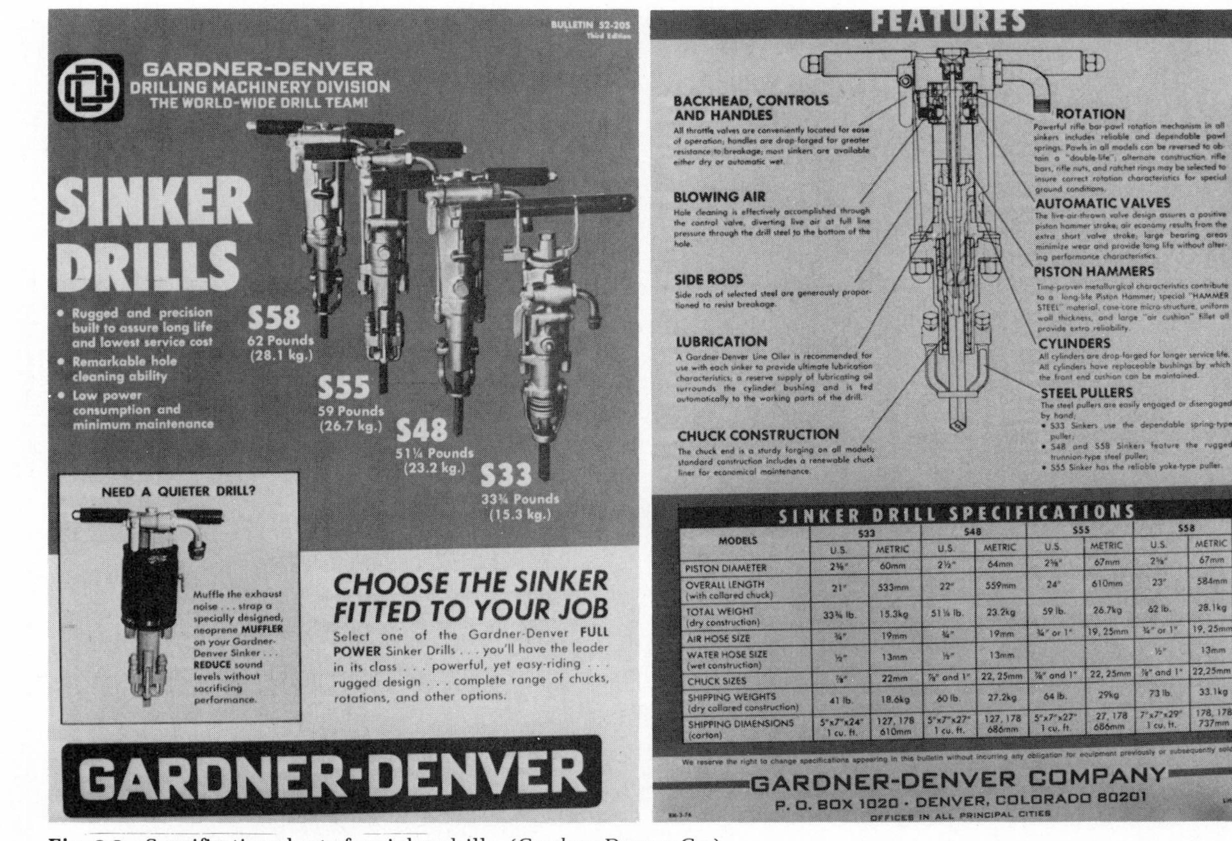

MODELS	S33		S48		S55		S58	
	U.S.	METRIC	U.S.	METRIC	U.S.	METRIC	U.S.	METRIC
PISTON DIAMETER	2⅜"	60mm	2½"	64mm	2⅝"	67mm	2⅝"	67mm
OVERALL LENGTH (with collared chuck)	21"	533mm	22"	559mm	24"	610mm	23"	584mm
TOTAL WEIGHT (dry construction)	33¾ lb.	15.3kg	51¼ lb.	23.2kg	59 lb.	26.7kg	62 lb.	28.1kg
AIR HOSE SIZE	¾"	19mm	¾"	19mm	¾" or 1"	19, 25mm	¾" or 1"	19, 25mm
WATER HOSE SIZE (wet construction)	½"	13mm	½"	13mm			½"	13mm
CHUCK SIZES	⅞"	22mm	⅞" and 1"	22, 25mm	⅞" and 1"	22, 25mm	⅞" and 1"	22,25mm
SHIPPING WEIGHTS (dry collared construction)	41 lb.	18.6kg	60 lb.	27.2kg	64 lb.	29kg	73 lb.	33.1kg
SHIPPING DIMENSIONS (carton)	5"x7"x24" 1 cu. ft.	127, 178 610mm	5"x7"x27" 1 cu. ft.	127, 178 686mm	5"x7"x27" 1 cu. ft.	27, 178 686mm	7"x7"x29" 1 cu. ft.	178, 178 737mm

Fig. 3.2 Specification sheets for sinker drills. (*Gardner-Denver Co.*)

Fig. 3.3 Row of crawler drills at work. (*Joy Manufacturing*)

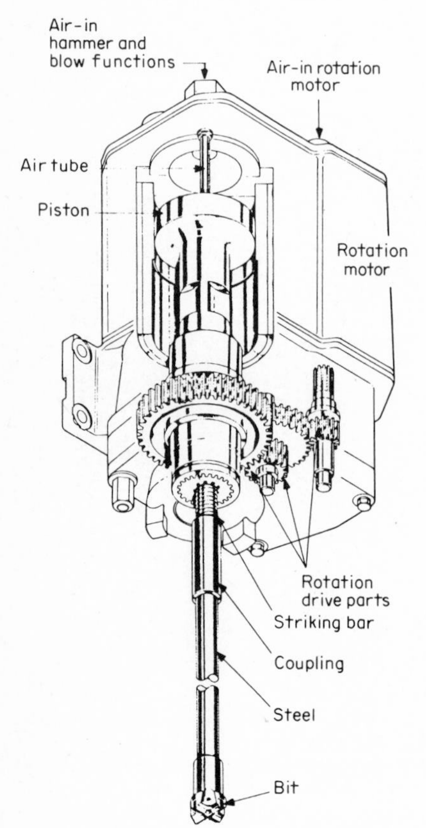

Fig. 3.4 Isometric sketch of an independent rotation drifter. (*Joy Manufacturing*)

creased or not used at all. Also, in hard rocks, where percussion is most needed, the hammer can be greater and the rotation slowed down.

Independent rotation is achieved by a gear motor operating independently of the piston. The motor is mounted on the top or side of the hammer. (See Figure 3.4.)

Because of its carriage, an external-percussion drill is well suited for very rough terrain. It can drill small-diameter holes (2 to 6 in, or 50 to 152 mm) in very hard rock. Also because the piston is larger than the bit (generally it should be at least 1 in, or 25 mm, larger), the drill's penetration rate for holes less than 60 ft (18 m) in hard rock is probably the most efficient for any drill.

External-percussion drills tend to be limited to shallower holes, and their inherent noise creates extra cost to make them comply with rising pollution standards. Because the drill steel and couplings transmit the percussive forces to the rock, the fatigue of the drill string is greater than with other drill systems, and the life of drill steel and couplings is shorter. (See Figures 3.5 and 3.6.)

External-percussion drills are usually mounted on a self-propelled crawler undercarriage. The crawler and boom are either air- or hydraulic-powered, the boom controls generally being hydraulic. The thrust comes from the weight of the drill rig and is applied by a chain and sprocket.

The drill rods for this type of drill are 1 to 2 in (25 to 50 mm) in diameter

Fig. 3.5 Changing drill steel on a crawler drill. (*Joy Manufacturing*)

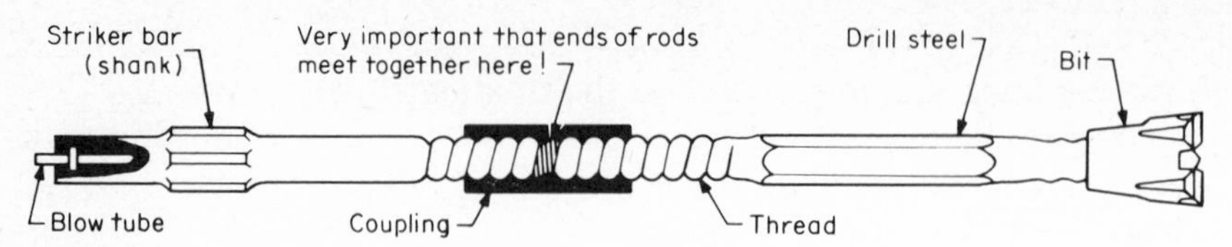

Fig. 3.6 The drill string. Energy is transmitted from the piston through the drill string to penetrate the rock. (*Worthington Corp.*)

and 8 to 12 ft (2.5 to 3.5 m) in length, and they have a hollow center for flushing with bits of either chisel or button characteristics.

Maintenance The maintenance requirements of a surface percussion drill are generally restricted to lubrication. The hammering is lubricated by a pressurized oil tank or bottle that automatically feeds an adjustable amount of oil to the hammer. Oil is transported to the hammer with the air: the air enters the oil bottle, where the oil is atomized in the air and passes to the hammer. With many oil bottles, when the bottle runs out of oil the air is shut off by a valve within the oil bottle, to prevent running the hammer dry and causing overheating. It is good procedure to turn the hammer on briefly before drilling to inspect the air exhaust to make sure it contains oil vapor indicating that the hammer is receiving lubricant.

DOWN-THE-HOLE DRILLS

Down-the-hole percussion drills have the drifter located behind the bit instead of mounted on a travel boom; the drifter goes into the hole with the bit. This offers the advantage of increased depth and larger boreholes. (See Figure 3.7.)

Down-the-hole drills are generally best suited for holes 4 to 9 in (101 to 230 mm) in diameter. The bit has to be at least ½ in (12 mm) in diameter larger than the drill because the drill goes into the borehole with the bit. The drills generally require 100 to 250 lb/in^2 (70,310 to 175,775 kg/m^2) of air with 4000 ft/min (1200 m/min) of velocity to flush the fines out of the holes, although the exhaust air provides some of the flushing. The down-the-hole drill requires 2000 to 4000 lb (900 to 1800 kg) of thrust at 10 to 100 r/min.

Down-the-hole drills are well suited for high penetration in deep holes, because, when the drill always remains with the bit, there is no power loss as there is with the surface drill, with its great length of drill steel (see Table 3.1). The noise of the drill is less, because the hammer, which generates the noise, is in the hole. Also, the drill rods have a longer life, because all they transmit is thrust, not percussion energy.

A major problem with this type of drill is in caving-type holes. If the bit becomes blocked in the hole and cannot be removed from the hole you lose the drill with it. The minimum hole that can be drilled is 4 in (100 mm) in diameter, and the penetration rate in shallow holes is slower than that of external drills, because the piston is smaller than the bit.

ROTARY DRILL

The rotary drill uses a combination of rotation and thrust to drill a hole in soft rocks and earth. It can drill holes with diameters of 6 to 9 in (150 to 230 mm) and is generally used in areas of soft to medium-hard drilling and easily accessible terrain.

The penetration of a rotary bit depends greatly on bit design, adaptability to the geological conditions, energy output of the drill, and applied thrust

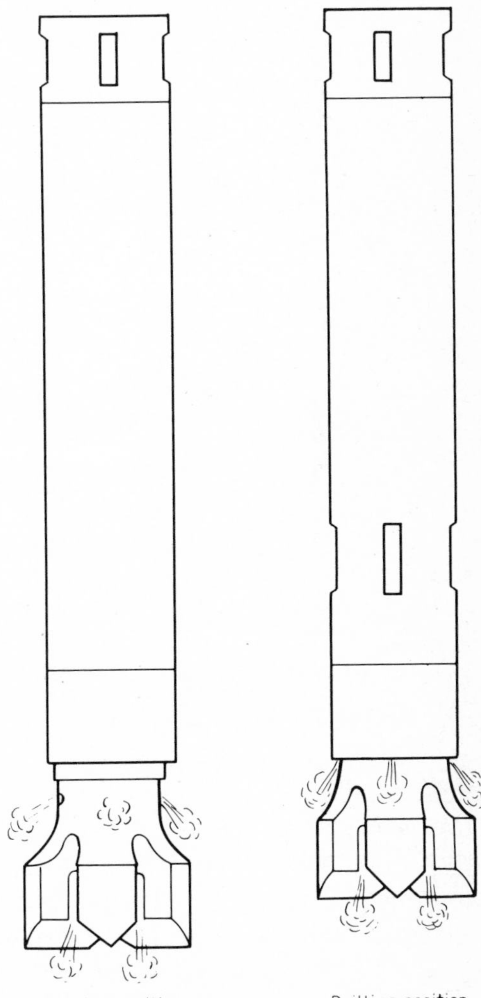

Fig. 3.7 A down-the-hole drill.

TABLE 3.1 Percentage of Loss on Addition of Drill Steel for External Drifter

Lengths of drill steel	Penetration rate loss, %
1	0
2	20
3	40
4 and 5	50
6 to 8	60

Steel beyond 8 is generally not used.

Example. Drilling a 40-ft hole, using a 10-ft drill steel, overall penetration rate = 80 ft/hr:

Steel no.	Penetration rate, ft/hr
1	80
2	64
3	48
4	40

Average penetration rate = 58 ft/h.

(down pressure). Also, the depth of cut and the rock strength affect the penetration. With the rotary drill the rock strength is described mostly in terms of shear strength, because the rotary acts mostly by shear stresses imposed on the rock. (See Table 3.2.)

The rotary drill is the most versatile drill for various rock types. In soft rock it employs a drag bit that uses a plowing action to impose shear stresses on the rock. In medium-hard rocks it uses a roller-type bit, i.e., a bit with heavy, mounted rollers, which when rotated create a gouging or compressive force on the rock, causing chipping. The rotary drill is generally used for hole diameters of 6 to 9 in (150 to 230 mm) and in 30- to 60-ft (9- to 18-m) cuts.

The drill system of a rotary drill consists of the power source, consisting of the downward thrust and the rotation system; the drill rod, which is hollow to function as a flushing medium; and the bit.

Main Components of a Rotary Drill

The main components of the rotary drill are the mounting, the power sources, the rotational system, the pulldown and hoisting system, the circulation system, and the mast and rod handling system.

Mounting The rotary drill can be mounted on a truck chassis or a crawler undercarriage. The choice of mounting depends on the intended use and the size of the drill. For a small rotary drill it may be most desirable to mount the drill on a truck chassis, not only for the mobility but also to permit the drill to be powered by the takeoff from the truck engine.

The need for mobility helps determine whether to use a truck- or a crawler-mounted drill. A truck-mounted drill obviously is more mobile, whereas a crawler-mounted one moves more slowly and, for any long distance, has to be loaded onto a truck and hauled. Although the crawler lacks the long-range mobility of a truck-mounted drill, it enjoys the advantage of being more mobile in rough terrain, being able to maneuver in terrains where trucks cannot.

Power Source The power source for rotary drills can be either gas, diesel, or electric. Gas would be more common with small drilling rigs, whereas diesel is generally the power source for the larger drills. Electric power is the most efficient, but its use is quite limited, because it not only restricts mobility but also requires a source of electricity, which is generally not found on construction sites. However, occasionally an open-pit mine or quarry has a layout conducive to electric power.

The power from the power source is transmitted by either mechanical or hydraulic means, usually the latter. Hydraulic power is used for tramming, for the leveling jacks, for the hoisting motors, for rotation, for the pulldown or thrust, and for running the air compressor.

The horsepower requirement for a particular drill is available from the manufacturer. However, the rotation motors require 60 to 140 hp; the hoisting and tramming require 35 to 75 hp; the thrust requires 20 hp; the air compressor requires 300 to 400 hp; and miscellaneous power users require another 40 to 50 hp. It must be clear that the total of these values is an overstatement, because not all these functions are being used at once.

TABLE 3.2 Rock Shear Strength

Rock	Shear strength, lb/in^2
Sandstone, soft	1,500
Sandstone, medium	3,050
Sandstone, hard gray	4,720
Sandstone, fine-grained brown	3,600
Sandstone, medium-grained friable gray	2,840
Limestone, hard fossiliferous	4,160
Limestone, hard gray	6,520
Limestone, very fossiliferous medium crystalline	7,600
Gneiss, gray granitoid	9,300
Siltstone, medium-hard sandy	3,000
Dolomite	12,700
Quartzite	10,600

Rotation The rotation for a rotary drill can be provided by a kelly bar system or by a top head drive powered by an electric or hydraulic motor.

Kelly Bar: The kelly bar uses a rotary table mechanically or hydraulically driven by the power source, or prime mover. The kelly bar system includes a square, hexagonal, or splined kelly bar attached to the drill pipe, which slides through a matching opening in the rotation table. As the table rotates the kelly bar and drill rod rotate with the table.

Unless the drill rod is shaped like the kelly bar, the kelly bar must be removed when additional steel is added as the hole progresses.

Top Head Drive: The top head drive is an electric or hydraulic gear-driven motor. The rotation is applied to the end of the drill rod, enabling the entire drive and drill string to move up and down the mast much as in an external-percussion drill.

Thrust The pulldown force is the thrust that the drilling applies to the bit. In general, rotary bits require 8000 lb of thrust per inch (1430 kg/cm) of bit diameter, depending on the rock characteristics. To determine the thrust required, multiply the diameter of the bit in inches by 8000. For example, a 7-in bit requires 7 × 8000 or 56,000 lb of downward thrust.

The downward thrust is achieved by using the weight of the drill, against which push either hydraulic pistons, cables, or a chain-and-sprocket combination similar to that used on external-percussion drills.

Flushing The circulating or flushing system requires approximately 30 to 60 lb/in^2 of air to flush the cuttings out of the borehole. This is achieved by forcing compressed air down the drill steel. Using the hollow center of the drill rod, the air passes down the steel and out through ports in the drill bit to blow the cuttings, or fines, up the borehole between the drill steel and the

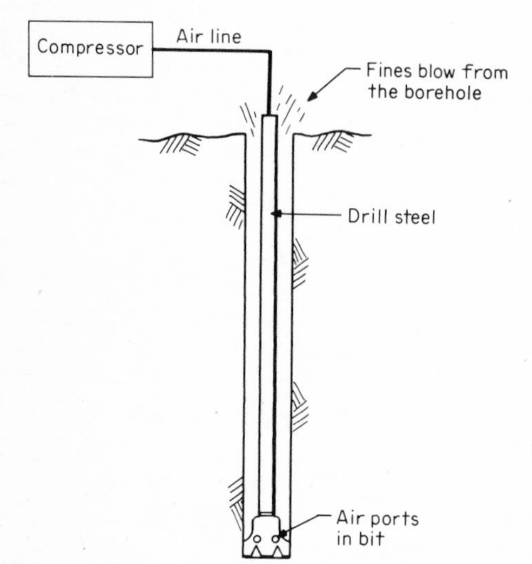

Fig. 3.8 The air flushing system.

walls of the borehole to the atmosphere or to a collection system attached to the drill rig. (See Figure 3.8.)

Mast Masts can be built to accommodate almost any drill steel length. However, lengths below 60 ft (18 m) are the most common. The masts are generally constructed in a truss pattern of tubular steel, much like crane booms. They are hinged and are connected to hydraulic cylinders that permit them to be lowered to a horizontal position for moving. Many have pinning capabilities that allow the mast to be pinned to permit drilling at angles as much as 30° out of vertical.

THE DRILL STRING

The drill steel can be of any circumferential shape: it may be round, square, hexagonal, or any shape that fits the specific requirement. All drill steels, or rods, transmit energy to the bit. The type and amount of energy needed determine the size, shape, and strength of the drill rod.

Drill Rod Failure

Drill rods for external percussion drills are subjected to the most energy, and therefore they have to be the strongest. This drill steel transmits thrust, percussion, and torque, or rotation. The most common cause of premature drill rod failure is underfeeding of the thrust. This causes heat, which creates a free end-reflection wave, causing additional fatigue in the drill rod, the couplings, the striking bar, and the hammer itself.

Another cause of drill string failure is misalignment. If the drill string is improperly aligned with the direction of the hole, a whipping action is created, causing additional heat and a flexural fatigue in the drill string. This will cause breakage in the steel, the coupling, and the striking bars. (See Figure 3.9.)

Lengths of drill steel should be rotated in their relative positions on the drill string, since each position on the drill string has different stresses. For example, if you are using three lengths of drill steel, each should serve as length no. 1 (the one next to the bit) only once for every three holes. By being rotated, each drill steel is exposed to changing stresses, instead of one concentrated stress, so that steel life is lengthened. However, be careful to match threads of equal wear: if new threads are used with older, worn threads, the new threads will wear out faster than if used with other new threads.

The best way to avoid misalignment is to make sure the drill is securely set up over the hole with the drill mounting weight resting on the boom, or mast, for stability.

When starting the hole be careful not to overfeed, because this can move the mast, causing misalignment. Also, when starting the hole, use a high rotation speed and a slow hammer in the overburden to make a clear entry for

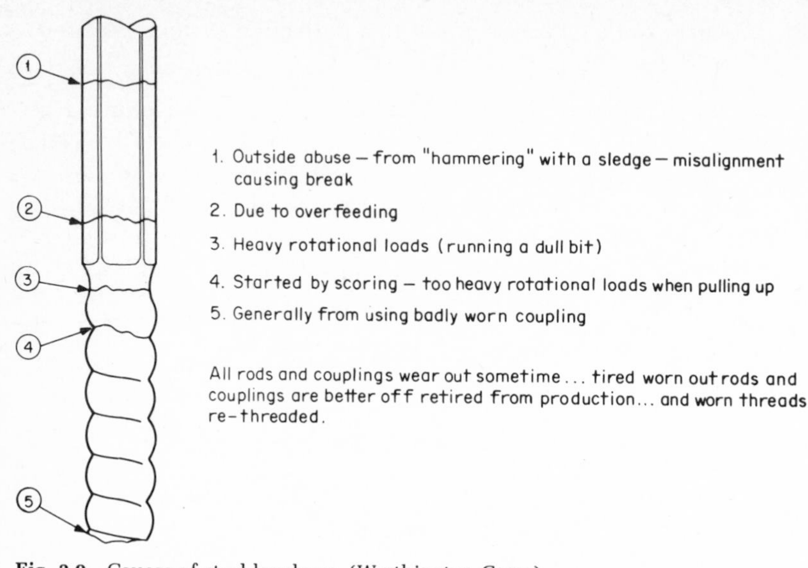

Fig. 3.9 Causes of steel breakage. (*Worthington Corp.*)

a good collar. If the collar, or entry to the hole, is rough, material may fall into the hole during drilling, causing the bit to become "hung" (unable to be pulled from the hole) or causing resistance to the rotation.

Striking Bar and Coupling Failure

Excessive rotation and overfeeding are the principal causes of striking bar and coupling failure. To ensure maximum life of the drill string, the feed and rotation must match the drilling condition. Too much rotation creates additional torque stresses on the drill string. Overfeed creates extreme vibration in the drill string, causing increased frequency of stress and thus reducing the life of the string.

Drill steel for down-the-hole and rotary drills is not subjected to the percussive energy of external-percussion drills. However, misalignment, overfeed, and excessive rotation can still decrease the life of the drill rod for these drills. Therefore, the same principle of matching the thrust and rotation to the conditions still applies. These drill rods should be 2 in (50 mm) smaller than the borehole. (See Figure 3.10.)

Bits

Bits for percussion drills are generally chisel or button bits. Chisel bits may take the form of a cross, an X, or a chisel. The cross consists of four tungsten carbide inserts at right angles to each other forming a cross. The X bit, like the cross bit, has four inserts at angles to each other forming an X. The chisel bit has one tungsten carbide insert centered in the form of a chisel. (See Figure 3.11.)

1. Worn chuck driver bushing
2. Overfeeding
3. Heavy rotation loads (running a dull bit)
4. Excessive scoring – too much rotation when pulling up

5 and 6. Misalignment

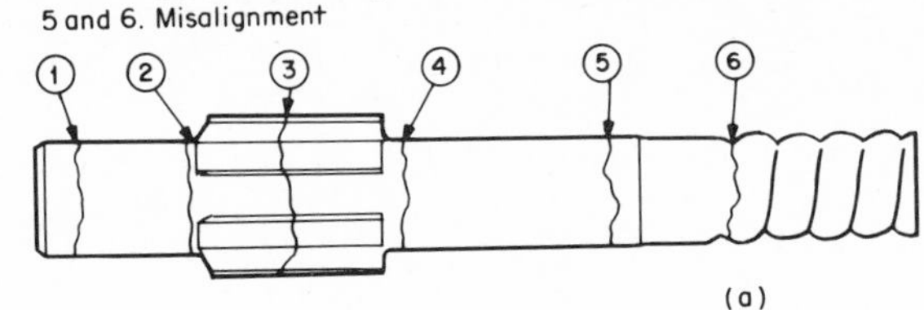

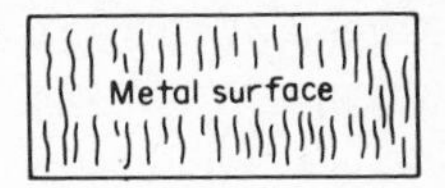

Typical scoring marks – Lack of lubrication and excessive pressure. These marks can be the start of cracks that will finally start breakage!

(a)

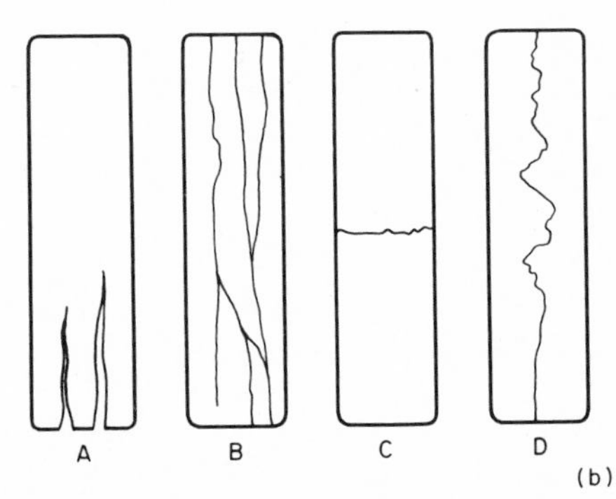

A. Excessive hammering without rotation or hammering when not "made uptight"

B. Outside abuse such as "hammering" with a sledge to loosen

C. Bending stresses – misalignment – ends of steels not "butted" together

D. Generally from over tightening of joint – too heavy rotation – running a drill bit

(b)

Fig. 3.10 Causes of (*a*) striker bar failures and (*b*) coupling failures. (*Worthington Corp.*)

Button bits are flat-bottomed and cylindrical, with cylindrical tungsten carbide inserts spaced about the bit. These bits are more common in sizes 2 to 6 in (50 to 152 mm) in diameter and are generally not sharpened but are used until they completely wear out.

Chisel bits may be ground or sharpened as the obtuse point on the carbide insert wears or dulls. Chisel bits should have a cutting angle of 110° with a

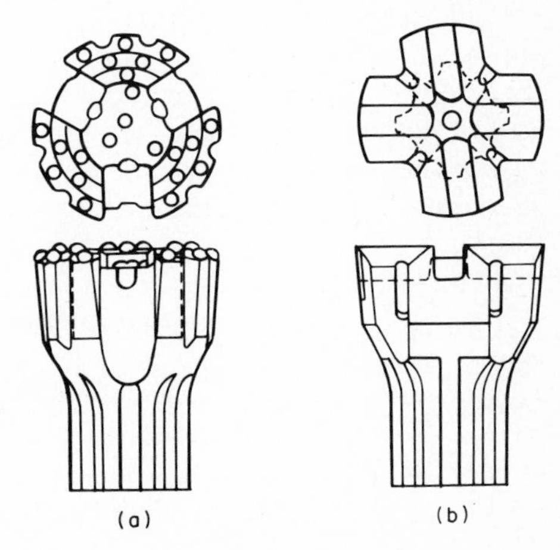

Fig. 3.11 Bits: (*a*) button and (*b*) X-type chisel.

gauge of 3° and at least 1/32 in of insert should remain beyond the bit housing. Gauging is the reduction in the protrusions of the inserts beyond the bit housing. The carbide inserts, because they are what cuts the rock, should be the wider portion of the bit; if they are not, the bit will bind in the hole.

When using more than one bit in a single hole you must make sure that the second bit is smaller than the first, and so on. If it is not smaller, the bit will bind in the hole and you will not be able to remove it.

Bits for Rotary Drills Rotary drills use either drag or roller bits. Drag bits use a plowing action and are best suited for soils, soft shales, and poorly cemented sandstones. The drag bit has six or eight fingers that are set to drag when the bit is rotated, so that they plow the material being drilled.

There are two basic types of drag bits: integral-blade and replaceable-cutter-element. Both types have cutter elements that are hard-faced with tungsten carbide inserts to resist wear, but with the replaceable-cutter bit, the blade or carbide inserts can be replaced when worn. The drag bit may be of the integral-blade type with two or three replacement fingers.

Rolling-cutter bits consist of two or three cones mounted with steel teeth or tungsten carbide inserts. The bicone bit is used primarily for tilted laminated formations because it tends to drift less than the tricone.

Roller bits can be used in material of any type or hardness. For soft material the steel teeth are widely spaced and mounted askew to the rolling altitude to cause the bit to drag when rotated, gouging or plowing the material. For hard formations the bit has small teeth in tungsten carbide inserts with the teeth at the proper rolling altitude so that the rock fails under the bit from indentation and shattering.

These bits require a regular or jet-type flushing system to clean the hole.

HYDRAULIC DRILLS

In the mid-1970s hydraulic drills were first introduced to mines in the United States. Since that time they have grown rapidly in popularity, and it appears that the use of hydraulic drills will increase even more in the 1980s. (See Figure 3.12.)

The reason for this increased interest in the hydraulic drill is that in most cases users are getting greater production from it at reduced energy costs. The hydraulic drill is particularly well suited for use in underground drilling because of the reduced noise levels and the reduction in fog produced by drill air exhaust (air is still used for flushing).

Users are also claiming that the drill steel and bits are lasting longer. It is understandable, when one considers what normally causes bits to wear. Bits are generally worn by rotation against the abrasive rock, not the percussion energy. Therefore, the hydraulic drill, with its many blows per minute, can greatly increase the penetration rate without decreasing bit life. Actually the bit life increases because there is more footage drilled between grindings.

The drill steel has greater life due to the nature of the stress distribution. The piston in a hydraulic drill is smaller in diameter and much longer than the piston in a hammer. The effect of this is that the stresses are in the form of a longer, more extended shock wave that covers a larger area of drill steel. Therefore, the stresses are distributed over more of the drill steel, lessening the effect on any one area of it and decreasing the life of the drill rod. (See Figure 3.13.)

The primary disadvantage of the hydraulic drill is the need for specialized repair. That is, job mechanics are not well enough trained and experienced to be efficient at repairing these drills. With an air hammer the mechanic's primary tools are a 2-lb hammer, a 24-in pipe wrench, and a spool of no. 9 wire. Some of these drills manage to be repaired with the crudest of tools and without much skill. However, the hydraulic hammer requires a knowledge of hydraulics and skill with tools. The problem, of course, will be solved with time. As drill "doctors" become familiar with the machinery, the maintenance problems will become smaller.

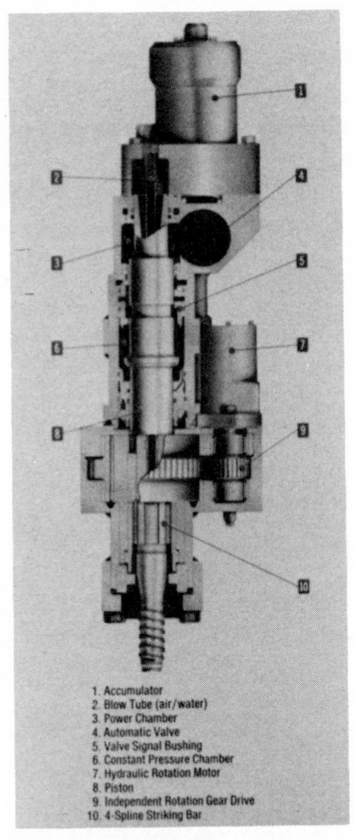

Fig. 3.12 Hydraulic drifter. (*LeRoi Division, Dresser Industries, Inc.*)

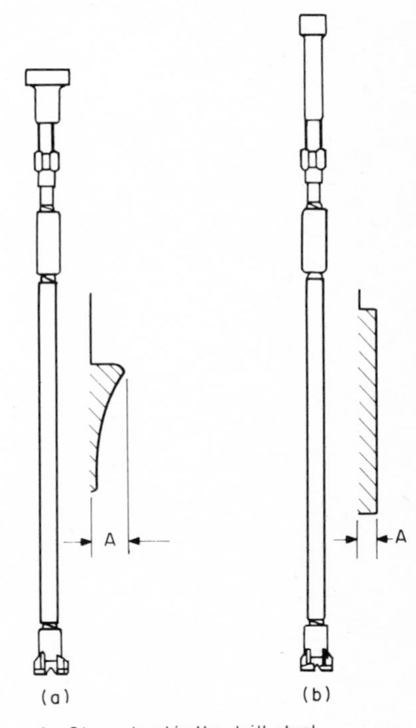

Fig. 3.13 Comparison of drill steel stress levels between (*a*) pneumatic rock drill and (*b*) hydraulic rock drill. (*Atlas Copco Corp.*)

No one would dispute that eventually, because of their efficiency, practically all hard-rock drilling will be done with hydraulic drills. However, this will take several years. At this writing, much development is required to offer the variety of drills that are offered in the air hammer. Also a contractor will not scrap an entire store of air drills to go hydraulic; the transition will be one unit at a time as each one needs replacing. It can be likened to the switch from cable to hydraulic "dozers": if you look you can still find cable dozers being used. Pneumatic drills are to be with us for a long time yet; however, hydraulic drills will see more and more use as time goes on.

DRILL SELECTION

The type of drill selected is dependent on a number of factors, economic, technical, and logistic. It is important, when selecting a drill for a particular job, to analyze the need according to the following criteria.

Size of Project

The size of a project requires attention to economic and logistic factors in addition to the obvious technical ones. Is the project large enough to make it feasible to use the technically most desired machine? Is the project so large that logistically the suppliers cannot provide enough of the desired drills? One has to check to see if there are enough drills available to do a job this size. At the other extreme, is the project so small that the logistics involved do not warrant bringing in the proper drill? For example, if a small job, say a swimming pool, has cuts ranging in depth from 8 to 10 ft (2.5 to 3 m), you would probably ordinarily use a small crawler drill. However, the horizontal dimensions may be only 12 ft by 12 ft (3.5 m by 3.5 m); thus, by the time a truck can pick up the smaller crawler drill and deliver it, the job can be drilled out with jackhammers. Therefore, the size of the project dictates what type of drill should be used.

Hole Diameter

The diameter of the hole will dictate what type of drill to use. For example, if in the blast design it is determined that a 4-in (100-mm) hole is required, that alone, even before anything else is known about the job, will prohibit the use of jackhammers and big rotary drills. Of course, more information is needed; the selection process involves analyzing each requirement before the final selection can be made. (See Table 3.3.)

Depth of Cut

The depth of cut further narrows the choice of drills. If the depth of cut for the 4-in (100-mm) hole is 90 ft (27 m), then in all likelihood, the down-the-hole drill is the best to use. However, what if the depth of cut is 60 ft (18 m)? It was stated earlier that 60 ft (18 m) is the deepest efficient depth for a

TABLE 3.3 Drilling Speed vs. Bit Size Conversion Chart

Bit size for known speed, in	Bit size to be used, in												
	5	4½	4	3½	3	2¾	2½	2¼	2	1⅞	1¾	1⅝	1½
5	1.000	1.170	1.400	1.710	2.150	2.460	2.830	3.310	3.960	4.350	4.820	5.410	6.100
4½	0.853	1.000	1.190	1.450	1.830	2.090	2.410	2.820	3.370	3.710	4.110	4.610	5.190
4	0.715	0.838	1.000	1.220	1.540	1.750	2.020	2.360	2.820	3.110	3.450	3.860	4.350
3½	0.585	0.688	0.820	1.000	1.260	1.440	1.650	1.940	2.320	2.550	2.820	3.170	3.560
3	0.464	0.545	0.650	0.793	1.000	1.140	1.310	1.540	1.840	2.020	2.240	2.510	2.820
2¾	0.407	0.478	0.570	0.695	0.878	1.000	1.150	1.350	1.610	1.770	1.970	2.200	2.480
2½	0.354	0.415	0.495	0.605	0.762	0.869	1.000	1.170	1.400	1.540	1.710	1.910	2.150
2¼	0.302	0.352	0.422	0.515	0.650	0.742	0.854	1.000	1.190	1.310	1.460	1.630	1.840
2	0.252	0.296	0.354	0.432	0.544	0.620	0.715	0.838	1.000	1.100	1.220	1.370	1.540
1⅞	0.229	0.279	0.321	0.392	0.494	0.564	0.650	0.760	0.910	1.000	1.110	1.240	1.400
1¾	0.207	0.243	0.290	0.354	0.446	0.509	0.586	0.686	0.821	0.904	1.000	1.120	1.260
1⅝	0.185	0.217	0.259	0.315	0.398	0.454	0.523	0.612	0.732	0.805	0.893	1.000	1.120
1½	0.164	0.193	0.230	0.280	0.354	0.403	0.464	0.544	0.650	0.717	0.792	0.889	1.000

When the drilling speed for a given bit size is known, the estimated drilling speed for another size can be found as follows:

1. Find the bit size for the known drilling speed in the *left-hand column*.
2. Follow the chart along this line until you reach the column under the size of the bit for which the estimated drilling speed is desired. This number is the conversion factor.
3. Multiply the known drilling speed by this conversion factor to obtain the estimated drilling speed for the bit size chosen.

Example: If the drilling speed for a 3-in bit = 24 in/min, for a 2½-in bit it will be:

$$1.31(24) = 31.4 \text{ in/min}$$

4. For metric bit sizes: in to mm = in × 25.4.

surface-mounted drill. In other words, there are certain gray areas of productivity where each contractor has to decide on the basis of experience which is the most productive method.

If the job is large enough, you can select one type of drill but also try another on the job to get an actual comparison. If you find that the drill you have chosen is not the best of the two, you may or may not be able to change at that time.

No matter what the best drill for the job is determined to be, if there is only one drill available, that is the drill to use.

Going back to the swimming pool example, if the only drill available had been the external-percussion drill, that is the drill that would have been used regardless of the problem of logistics and even if the average depth of cut were only 3 ft (0.9 m).

The textbook rule is that one uses only the drill best suited. However, the reality of the situation is that if no drills are available one goes in and does the job with pavement breakers. The point is to get the job done.

Rock Hardness

Another consideration is the rock hardness. This matters only in cases where this is a decision to be made between a percussion and a rotary drill. The contractor does not face these alternatives too often, but it is a consideration on large projects that permit the use of a rotary drill. (See Table 3.4.)

Capital

A contractor who has determined the best drill for a job then has to see if it can be afforded. Thus, if the drill does not fit into the budget, the contractor has to pick the best-suited affordable drill. Many larger jobs will yield enough income to warrant buying the desired machine. Also, one can generally rent or lease the machine from an equipment house or another contractor. Thus lack of capital is not a big problem for large contractors, but to the small contractor it is.

Cost

The most common factor in competitive drilling is the cost. Not only is the selection of the type of drill affected, but also of the manufacturer. The basic cost consideration is the cost per foot of borehole. That is, is the drill chosen the one that can drill the holes the cheapest? This is best analyzed on a production-versus-cost basis: the total hourly cost for each unit must be determined and then compared to the hourly production of each unit. This cost must also be compared to more empirical data; if the cost per foot of borehole is low but the service record of the supplier is poor, it may be prudent to acquire equipment from another source. It is imperative that the contractor maintain an awareness of the supplier's reliability. If the machine breaks down and the supplier cannot provide the repair, or at least the

TABLE 3.4 Hardness Chart (Mohs' Scale) for Various Minerals and Rocks

Mineral or rock	Hardness		Scratch test
Diamond	10.0		
Carborundum	9.5		
Sapphire	9.0		
Chrysoberyl	8.5		
Topaz	8.0		
Zircon	7.5		
Quartzite	7.0		
Chert	6.5	Percussion drilling	
Trap rock	6.0	Percussion drilling	
Magnetite	5.5	Percussion drilling	
Schist	5.0	Percussion drilling	Knife
Apatite	4.5	Percussion drilling	Knife
Granite	4.0	Percussion drilling	Knife
Dolomite	3.5	Rotary drilling	Knife
Limestone	3.0	Rotary drilling	Copper coin
Galena	2.5	Rotary drilling	Copper coin
Potash	2.0	Rotary drilling	Fingernail
Gypsum	1.5		Fingernail
Talc	1.0		Fingernail

NOTE: Hardness is the resistance of a *smooth* plane surface to abrasion and is measured in terms of the above scale. Approximations may be reached by scratching with fingernail, copper coin, and knife.

part to repair the machine quickly, the downtime on the project will certainly increase the cost of production.

After examining the equipment in light of the arguments presented, the production person, or persons, should be able to make the decision; i.e., accountants should not be given the task of selecting equipment. The equipment selection should be made, in light of information provided by the accountants, by the people charged with completing the project.

FOUR

COMPRESSORS

BOYLE'S LAW

Compressors are important in drilling and blasting because they provide the air to power the pneumatic drills. Compressors used for construction are generally portable and are called "positive-displacement compressors." That is, successive volumes of air are confined and then the volume is decreased, which in turn increases the pressure, according to Boyle's law:

$$P_1V_1 = P_2V_2$$

where P = pressure, generally in pounds per square inch (lb/in²)
V = volume, generally in cubic feet (ft³)

Thus, as the volume decreases the pressure increases. For example:

$$P_1V_1 = P_2V_2$$

means

$$\frac{P_1}{P_2} = \frac{V_2}{V_1}$$

If

$$P_1 = 100 \text{ lb/in}^2$$
$$V_1 = 2 \text{ ft}^3$$
$$V_2 = 1.5 \text{ ft}^3$$

then

$$\frac{100}{P_2} = \frac{1.5}{2}$$

$$P_2 = \frac{200}{1.5}$$
$$= 133.3 \text{ lb/in}^2$$

TYPES OF COMPRESSORS

There are primarily three types of compressors used in construction: (1) the reciprocating (piston), (2) the sliding-vane, and (3) the helical (rotary-screw).

The Piston Compressor

The piston compressor is single-acting when a single cylinder compresses the air on one side of the piston; it is double-acting when two single-acting pistons are operating in parallel in one casting. The piston compressor has both intake and discharge valves to control the flow of air in and out of the cylinder. These valves are spring-type and open by changes in pressure with the cylinder.

Cycles There are four cycles to a piston compressor: (1) compression, (2) discharge, (3) expansion, and (4) intake. In the compression portion of the cycle the air is compressed before it is released. The discharge portion is the point where the pressure in the cylinder is great enough that the discharge valve opens and releases the compressed air. In the expansion phase the piston is on the return but the intake valve has not yet opened. At this point

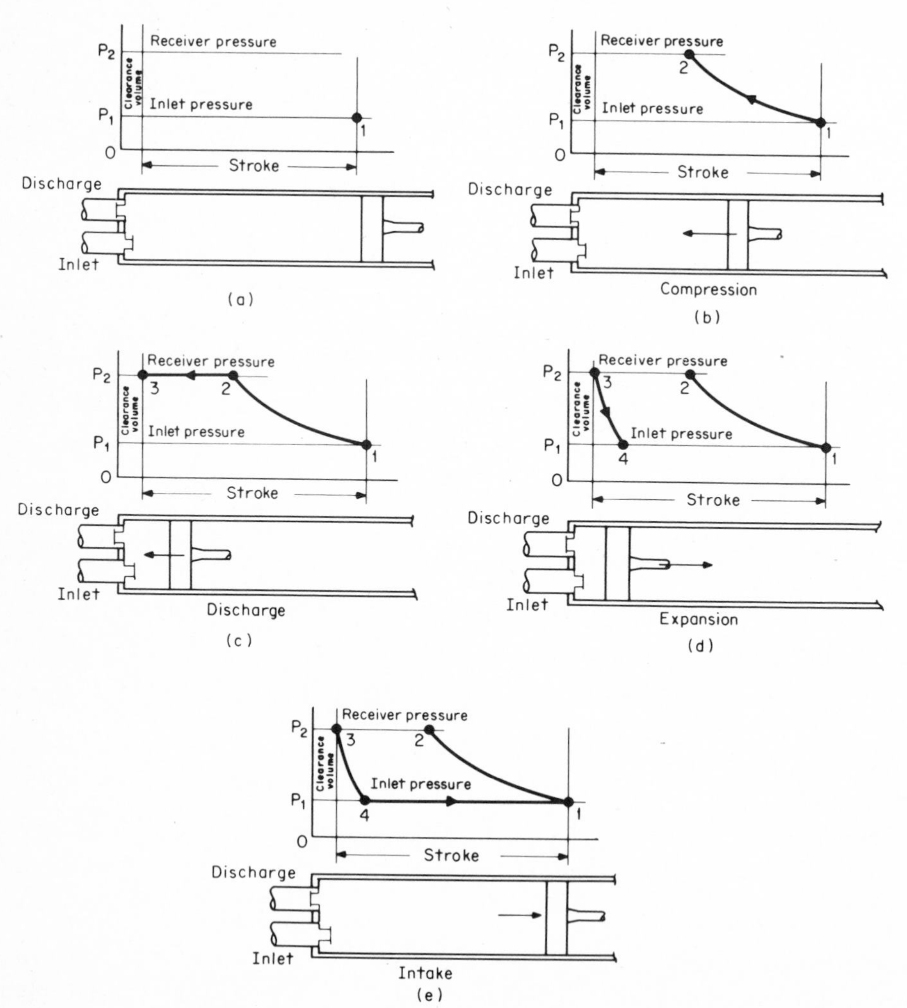

Fig. 4.1 The various steps in a reciprocating compressor cycle. (*Ingersoll-Rand Corp.*)

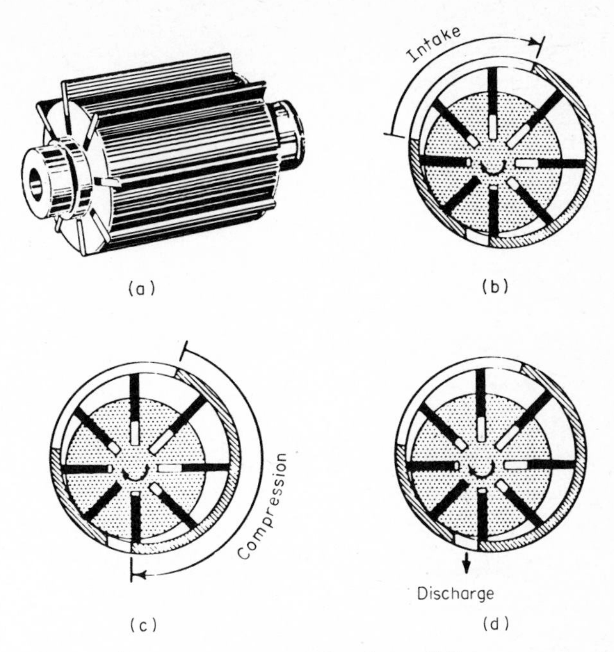

Fig. 4.2 The steps in compression for a sliding-vane rotary compressor: (*a*) rotor with nonmetallic sliding vanes; (*b*) as rotor turns, gas is trapped in pockets formed by vanes; (*c*) gas is gradually compressed as pockets become smaller; (*d*) compressed gas is pushed out through discharge port. (*Ingersoll-Rand Corp.*)

the air that was still entrapped in the cylinder after the release is expanded in volume, decreasing the pressure in the cylinder to a vacuum. When this vacuum is achieved, the intake valve will open, drawing air into the cylinder until the piston has reached the other end of the stroke. This intake phase completes the cycle, and the compression phase is due to begin again. (See Figure 4.1.)

In two-stage piston compressors the cylinders are proportioned to this compression ratio with the second stage smaller than the first because the air it receives is already partially compressed.

The Sliding-Vane Compressor

The rotary sliding-vane compressor consists of a cylindrical form, or casting, with slots about its circumference running longitudinally that contain sliding strips, or vanes. The casting rotates about a longitudinal axis within a housing that is eccentric about the casting. (See Figure 4.2.)

The vane compressor has no valves. Instead, the inlet and discharge of air depend on ports in the housing, which the vanes pass over as the casting rotates. These ports are so located that the inlet ports open into the pocket (or area) between the vanes at its widest, permitting the greatest possible volume of air to be trapped within the pocket. As the casting continues to rotate, the pocket travels through the section in which the distance between the edge of the casting and the housing wall decreases. This shorter distance, of course, reduces the volume of the pocket, compressing the air trapped within the pocket, and causes the vanes to slide farther into the slots. The column continues to decrease, increasing the air pressure, until the lead vane for the pocket passes a discharge port, releasing the entrapped, now compressed, air.

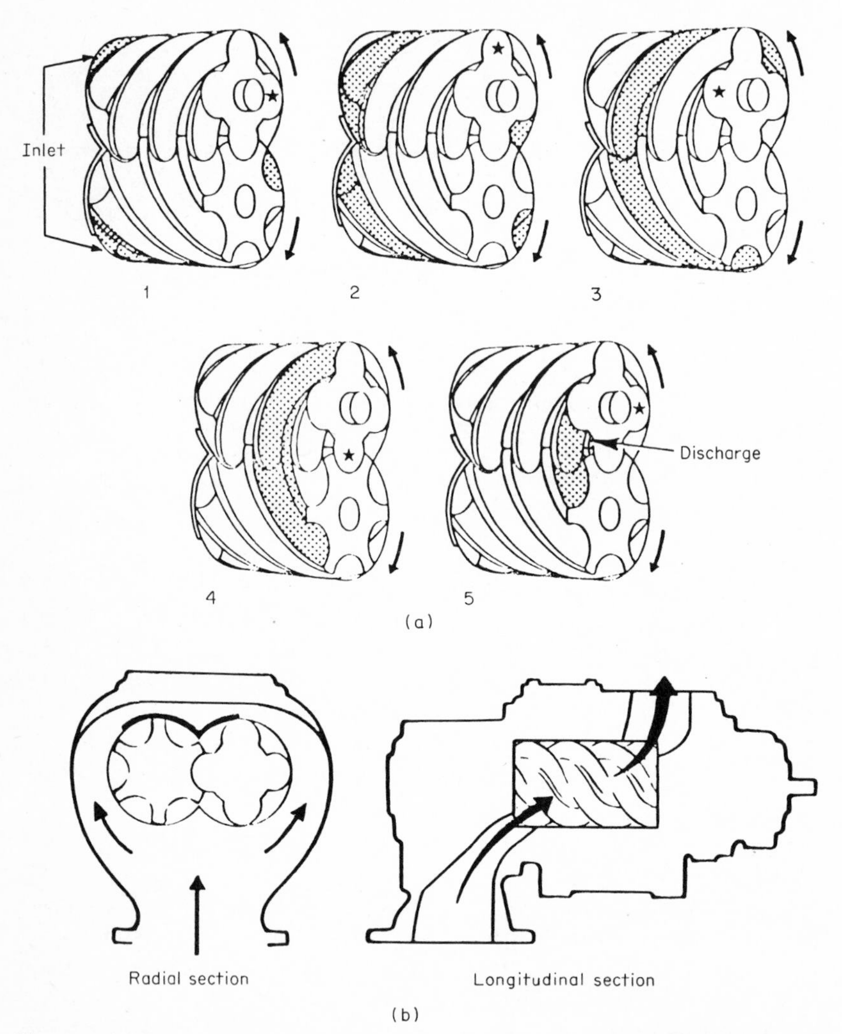

Fig. 4.3 (*a*) Steps in compression by a typical helical-lobe rotary compressor and (*b*) illustrative sections of a typical helical-lobe rotary compressor. (*Ingersoll-Rand Corp.*)

This is a continuous compression system; i.e., the inlet compression and discharge are always taking place, provided the casting is rotating. This method of compressing air is good for compressors up to 600 cubic feet per minute (ft^3/min).

The Helical (or Rotary-Screw) Compressor

The positive-displacement rotary unit compresses air between the intermeshing helical lobes, on the two screws, and the chambers of the housing. The most common type of screw compressor uses a flood of oil; i.e., the screws are flooded with oil, which acts as a seal, a cooler, and a lubricant. The screws may or may not be controlled by timing gears. (See Figure 4.3.)

The screw compressor compresses by internal compression, and the compression ratio is determined by the location of the opening edges of the discharge port and the angle of the screw. The main, or male, screw has fewer lobes and rotates faster.

As the air enters the intake port it is trapped between the lobes of the two screws and the housing. While the screws rotate the air is moved farther into the screw. As the air is being moved the spacing between the lobes and the flutes reduces, decreasing the volume and therefore compressing the air.

Screw compressors for construction have a range of 20 to 2000 ft^3/min.

With today's environmental regulations all construction compressors, whether piston, vane, or screw, come enclosed in sound-reducing shells to comply with noise-pollution levels.

PORTABILITY OF COMPRESSORS

Construction compressors are generally portable because of the mobility required. Because of this requirement, the drive power for the compressor is produced by an internal combustion engine. These engines are generally gasoline for smaller compressors (less than 150 ft^3/min) and diesel for the larger compressors. (See Figures 4.4. and 4.5.)

Cooling

Portable compressors use oil for cooling and recirculate it to lubricate and seal the vanes and screws and to remove heat.

The lubricating system contains an oil pump to regulate the oil going into the compressor. As the oil flows through the compressor it passes out the discharge port with the compressed air into the air receiver. In the air receiver there is an oil separator that separates most of the oil from the compressed air, dropping it to the bottom of the receiver tank. The oil is then piped through the oil cooler and filtered back to the pump inlet to be recirculated. (See Figure 4.6.)

The chief advantage to an oil-cooled system is that the temperature of the compressor and the discharged air will run 100 to 150°F (38 to 66°C) cooler

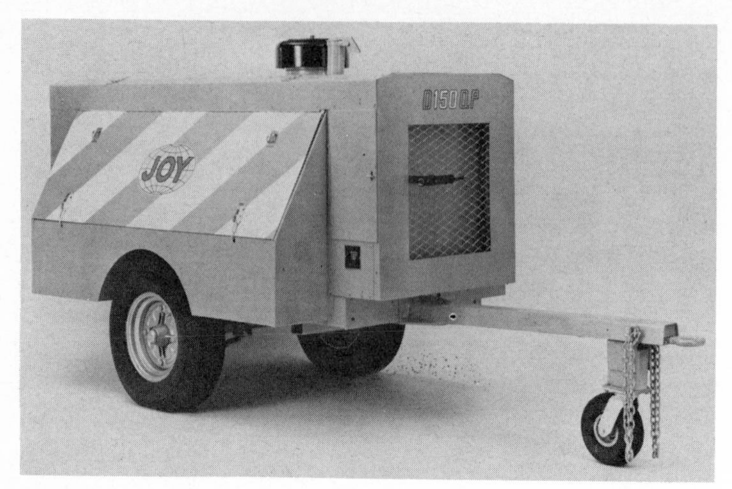

Fig. 4.4 A 150 ft^3/min portable compressor, commonplace on most construction projects. (*Joy Manufacturing.*)

than in a reciprocating compressor. This cooler operating temperature will, of course, increase the life of the machine.

Maintenance

The maintenance of a compressor, as with any equipment, should be at least to the minimum standards recommended by the manufacturers. As a general rule one would be hard-pressed to find someone who did too much maintenance. Even with the rising costs of energy, lubricating oil and grease are probably the least expensive items involved in compressing air, and yet probably the most neglected. Compressors, like any equipment at any price, operate better and longer with good lubrication and no dirt. Dirt is the nemesis of all equipment. It can plug fuel lines and oil lines, and if it gets

Fig. 4.5 A 600 ft^3/min compressor mounted on wheels for mobility in construction situations. (*Joy Manufacturing.*)

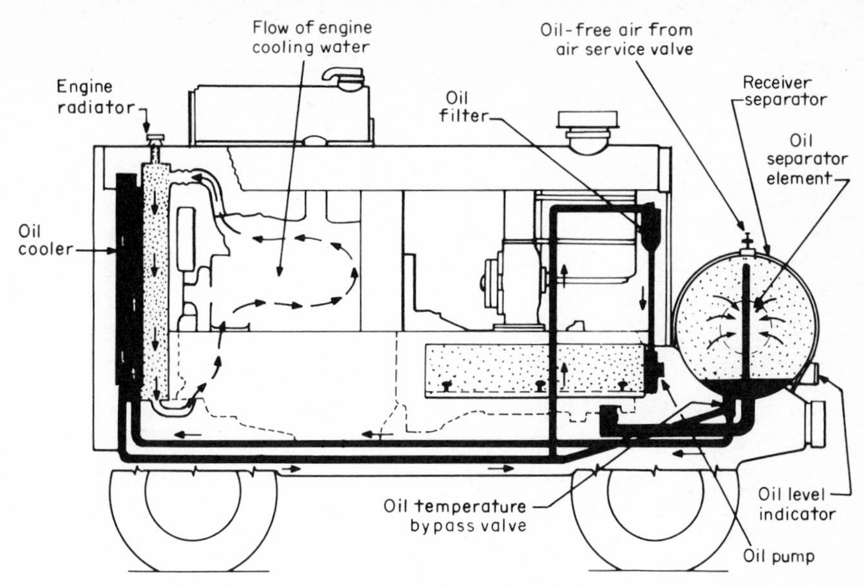

Fig. 4.6 Typical cooling-system circuits of an oil-cooled compressor and its engine drive. (*Ingersoll-Rand Corp.*)

into the compressor it can ruin the housing and internal parts, reducing the life of the equipment.

The best way to keep dirt out of the equipment is to minimize the dirt that the machine is exposed to and keep the air filters cleaned and changed often. In the extremely dusty situations that are often found around rock work, it may be necessary to change the filters as often as every shift.

Dust can have an effect on the temperature of the compressor if it is allowed to build up around the radiator and the oil cooler, restricting the air flow around the core. If the machine can't be kept away from the dust, it is imperative to clean the radiator by using an air hose and blowing back through the core of the radiator and the oil cooler. Thus the dust can be removed, so that the air can flow through the core, cooling the engine and the compressor. (See Figure 4.7.)

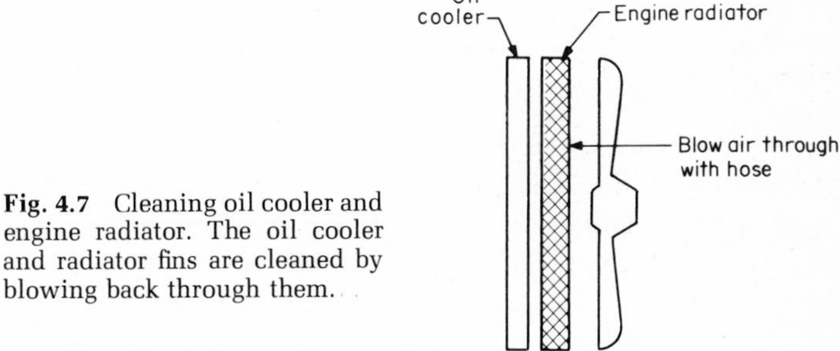

Fig. 4.7 Cleaning oil cooler and engine radiator. The oil cooler and radiator fins are cleaned by blowing back through them.

Oil[1]

There are compressor lubricating oils for different operating conditions. The oil required will vary with the temperature and frequency of use.

Motor or crankcase oil of grade SAE 10W may be used under certain operating conditions for rotary portable air compressors. The minimum requirement for compressor oil is that the oil comply with Specification MIL-2104B, grade SAE 10W. Any further qualification is extra assurance of the performance, but the aforementioned specification is the minimum requirement for compressor oil.

Motor oil may not be the recommended choice under more difficult operating conditions. For example:

1. Under conditions of light-duty operation and high relative humidity with vane- and screw-type compressors the condensed water would not boil off because of cool running conditions. Because of the detergent agents in the oil the condensation that built up would tend to become suspended, causing excessive slot wear in vane compressors and bearing wear in screw compressors. The best type of oil to use under these conditions is turbine-type oil.
2. For vane or screw compressors operating under continuously low ambient temperatures or with occasional temperatures of −20° to −40°F (−29 to −40°C), use automatic transmission fluid.
3. Also use automatic transmission fluid when vane- and screw-type compressors are operating under high temperatures, and in heavy-duty use, where oxidation of the oil, creating lacquers and varnishes, is a problem.

The compressor oil filter will be of the edge or replacement-element type. For servicing, the edge should be rotated every shift and the shell should be removed and cleaned approximately every 100 hours (h), but more often under extreme conditions. The element should be changed with each oil change, which should occur every 500–1000 h, depending on ambient and operating conditions.

To help relieve dirt and water problems it is best to drain the receiver-separator, fuel tanks, and fuel storage tanks daily, preferably before starting, to remove water and the dirt that has settled in the bottom of the tanks.

COMPRESSOR SELECTION

Before one can properly select the air source, the requirements must be determined. That is, an analysis of the air draw and air losses must be made. To make this analysis one must first determine the air-consuming equipment that is to be used and the method of conveying, or distributing, the air to the

[1] Oil recommendations from compressed air and gas data by Ingersoll-Rand.

equipment. The air-distributing system has to be analyzed to allow for friction losses in pipes, hoses, and fittings.

For example, assume that we have five drills, each requiring 800 ft^3/min of air at 100 psig for drilling and blowing. Also, let's assume that because of the difficult terrain and the size of the job the air has to be banked; that is, all the compressors have to be fed to one distribution center. Now, the amount of pipeline required to feed the drills is 700 ft (213 m), and each drill is attached to the header at the end of the pipeline by 150 ft (45.7 m) of 2½-in (63-mm) hose.

Ordinarily the first thought would be to multiply the air requirement of each drill by the number of drills to determine the total air requirement. Doing that, we obtain a total air requirement of 4000 ft^3/min. However, realizing that the pressure drops because of the friction in the distribution section, we elect to start our system with 4800 ft^3/min at 110 psig (77,341 kg/m^2) (four 1200 ft^3/min compressors).

Using Tables 4.1 to 4.5 we can determine the pressure losses due to friction in the system. Assuming a 6-in (152-mm) diameter pipe, we see that 4800 ft^3/min at 110 psig is not on Table 4.2. Therefore we have to interpolate. Interpolating between the values, arrive at a line pressure drop per 100 ft (30 m) of length of 0.48 psig. For the total pipeline we calculate:

$$n \times p = P$$

where p = pressure loss per length
P = total pressure loss
n = hundreds of feet of pipe

Thus,

$$7 \times 0.48 = 3.36 \text{ psig}$$

Therefore, at the distribution header there is a total availability of air for the drills of 4800 ft^3/min at 106.64 psig (110 − 3.36).

TABLE 4.1 Temperature Friction Loss

Flowing air temperature, °F (°C)	Friction loss multiplier
70(21)	1.00
130(54)	1.11
200(93)	1.24
250(121)	1.34
300(148)	1.43
350(176)	1.53

NOTE: Tables 4.2 to 4.5 are based on an air temperature of 70°. However, often the air temperature is much higher, creating even greater friction loss. This table gives multipliers for various air temperatures.

SOURCE: C. W. Gibbs, "New Compressed Air and Gas Data," Ingersoll-Rand Co., Phillipsburg, N.J., 1971, p. 34-161.

TABLE 4.2 Pressure Loss in Pounds for Each 100 ft of Straight Pipe

a. Nominal pipe size 3 in, Schedule 40

Free air, ft³/min	Line pressure, psig													
	10	15	20	30	40	50	75	100	125	150	200	250	300	350
300	.36	.30	.26	.20	.16	.13	.10							
500	.94	.78	.67	.52	.43	.36	.26	.20	.17	.14	.11			
750		1.69	1.44	1.12	.92	.78	.56	.44	.36	.30	.23	.19	.16	.14
1,000			2.50	1.94	1.59	1.34	.97	.76	.62	.53	.41	.33	.28	.24
1,500				4.30	3.52	2.98	2.15	1.68	1.38	1.17	.90	.73	.61	.53
2,000						5.29	3.81	2.99	2.47	2.08	1.60	1.30	1.09	.94
2,500							5.96	4.67	3.83	3.26	2.50	2.02	1.70	1.47
3,000							8.58	6.71	5.51	4.68	3.58	2.91	2.45	2.11
3,500								9.15	7.50	6.37	4.89	3.96	3.34	2.88
4,000								11.9	9.80	8.31	6.36	5.16	4.35	3.76
4,500									12.4	10.5	8.06	6.55	5.50	4.75
5,000									15.3	13.0	9.95	8.07	6.80	5.86
6,000											14.3	11.6	9.78	8.45
7,000											19.5	15.9	13.4	11.5

b. Nominal pipe size 4 in, Schedule 40

Free air, ft³/min	Line pressure, psig													
	10	15	20	30	40	50	75	100	125	150	200	250	300	350
500	.24	.20	.17	.13	.11									
750	.52	.43	.37	.29	.23	.20	.14	.11						
1,000	.90	.75	.64	.50	.41	.34	.25	.19	.16	.14	.10			
1,500		1.64	1.41	1.09	.89	.75	.54	.43	.35	.30	.23	.18	.16	.13

2,000			2.46	1.91	1.56	1.32	.95	.75	.61	.52	.40	.32	.27	.23
2,500				2.96	2.42	2.04	1.47	1.16	.95	.80	.62	.50	.42	.36
3,000				4.20	3.44	2.91	2.10	1.64	1.35	1.14	.88	.71	.60	.52
4,000						5.15	3.11	2.90	2.38	2.02	1.55	1.26	1.06	.91
5,000							5.75	4.50	3.70	3.14	2.40	1.95	1.64	1.42
6,000							8.22	6.45	5.30	4.50	3.44	2.79	2.35	2.03
7,000								8.77	7.20	6.10	4.68	3.80	3.20	2.76
8,000								11.5	9.40	8.00	6.11	4.95	4.17	3.60
10,000									14.7	12.5	9.55	7.75	6.52	5.62
12,000											13.8	11.2	9.40	8.10
14,000											18.8	15.2	12.8	11.0

c. Nominal pipe size 5 in, Schedule 40

Free air, ft³/min	Line pressure, psig													
	10	15	20	25	30	35	40	50	60	80	100	125	150	200
1,000	.29	.24	.20	.18	.16	.14	.13	.11						
1,500	.62	.52	.44	.39	.34	.31	.28	.24	.21	.16	.13	.11		
2,000	1.09	.91	.78	.68	.60	.54	.49	.42	.36	.28	.24	.19	.16	.13
2,500		1.39	1.19	1.04	.92	.83	.75	.64	.55	.44	.36	.30	.25	.19
3,000		1.98	1.69	1.48	1.31	1.18	1.07	.91	.79	.62	.51	.42	.36	.27
4,000			2.99	2.61	2.32	2.08	1.89	1.60	1.39	1.09	.90	.74	.63	.48
5,000				4.02	3.68	3.22	2.92	2.47	2.14	1.69	1.40	1.15	.97	.75
6,000						4.60	4.18	3.64	3.07	2.42	2.00	1.64	1.39	1.07
8,000								6.24	5.40	4.26	3.53	2.90	2.46	1.88
10,000									8.40	6.62	5.47	4.50	3.82	2.92
12,000										9.50	7.88	6.47	5.50	4.20
14,000											10.7	8.80	7.46	5.72
16,000												11.5	9.75	7.47
18,000												14.5	12.4	9.45
20,000													15.2	11.7

TABLE 4.2 Pressure Loss in Pounds for Each 100 ft of Straight Pipe (*Continued*)

d. Nominal pipe size 6 in, Schedule 40

Free air, ft³/min	Line pressure, psig													
	10	15	20	25	30	35	40	50	60	80	100	125	150	200
1,500	.25	.20	.18	.15	.14	.12	.11							
2,000	.43	.36	.31	.27	.24	.21	.19	.16	.14	.11				
2,500	.66	.55	.47	.41	.36	.33	.30	.25	.22	.17	.14	.12	.10	
3,000	.94	.78	.67	.58	.52	.47	.42	.36	.31	.24	.20	.17	.14	.11
4,000		1.36	1.16	1.01	.90	.81	.74	.62	.54	.43	.35	.29	.25	.19
5,000			1.80	1.58	1.40	1.26	1.14	.97	.84	.66	.55	.45	.38	.29
6,000			2.58	2.26	2.00	1.80	1.64	1.38	1.20	.95	.78	.64	.55	.42
8,000				3.98	3.54	3.18	2.89	2.44	2.12	1.67	1.38	1.13	.96	.74
10,000						4.90	4.45	3.77	3.26	2.57	2.13	1.75	1.48	1.14
12,500								5.85	5.06	4.00	3.30	2.71	2.30	1.76
15,000									7.30	5.75	4.75	3.90	3.32	2.54
17,500										7.82	6.47	5.30	4.50	3.46
20,000											8.45	6.95	5.88	4.51
25,000												10.8	9.20	7.05
30,000													13.3	10.2

e. Nominal pipe size 8 in, Schedule 40

Free air, ft³/min	Line pressure, psig													
	10	15	20	25	30	40	50	60	70	80	90	100	110	125
3,000	.23	.19	.17	.14	.13	.10								
4,000	.40	.34	.29	.25	.22	.18	.15	.13	.12	.11				
5,000	.57	.47	.41	.35	.31	.26	.22	.19	.17	.15	.13	.12	.11	.10

Free air, ft³/min	10	15	20	25	30	40	50	60	70	80	90	100	110	125
7,500	1.37	1.14	.98	.85	.76	.62	.52	.45	.40	.36	.32	.30	.27	.24
10,000		2.00	1.71	1.50	1.33	1.08	.92	.80	.70	.63	.57	.52	.48	.43
12,500			2.62	2.29	2.04	1.66	1.41	1.22	1.08	.96	.87	.80	.73	.65
15,000				3.30	2.94	2.39	2.02	1.75	1.55	1.38	1.25	1.14	1.06	.94
17,500					3.98	3.26	2.76	2.39	2.10	1.88	1.71	1.56	1.43	1.28
20,000						4.22	3.56	3.09	2.73	2.44	2.21	2.02	1.85	1.66
25,000							5.58	4.83	4.26	3.81	3.45	3.16	2.90	2.59
30,000								6.90	6.08	5.45	4.92	4.50	4.15	3.69
35,000									8.30	7.40	6.70	6.12	5.64	5.03
40,000										9.66	8.74	8.00	7.35	6.55
45,000											11.1	10.2	9.35	8.33

f. Nominal pipe size 10 in, Schedule 40

Free air, ft³/min	Line pressure, psig													
	10	15	20	25	30	40	50	60	70	80	90	100	110	125
5,000	.20	.16	.14	.12	.11									
7,500	.43	.36	.31	.27	.24	.19	.16	.14	.13	.11				
10,000	.75	.63	.54	.47	.42	.34	.29	.25	.22	.20	.18	.16	.15	.13
12,500	1.16	.97	.83	.72	.64	.52	.44	.38	.34	.30	.27	.25	.23	.21
15,000		1.38	1.18	1.03	.91	.75	.63	.55	.48	.43	.39	.36	.33	.29
17,500		1.87	1.60	1.40	1.24	1.01	.86	.74	.66	.59	.53	.49	.45	.40
20,000			2.07	1.81	1.60	1.31	1.11	.96	.85	.76	.69	.63	.58	.50
25,000				2.82	2.50	2.05	1.73	1.50	1.32	1.18	1.07	.98	.90	.80
30,000					3.58	2.92	2.47	2.14	1.89	1.69	1.53	1.40	1.28	1.15
35,000						3.98	3.37	2.92	2.58	2.30	2.08	1.90	1.75	1.56
40,000						5.20	4.40	3.80	3.55	3.00	2.72	2.48	2.28	2.04
50,000								5.90	5.20	4.65	4.20	3.85	3.54	3.16
60,000									7.50	6.70	6.07	5.55	5.10	4.55
70,000										9.13	8.25	7.55	6.95	6.20
80,000											10.8	9.85	9.05	8.10

TABLE 4.2 Pressure Loss in Pounds for Each 100 ft of Straight Pipe (*Continued*)

g. Nominal pipe size 12 in, Schedule 40

Free air, ft³/min	Line pressure, psig													
	10	12.5	15	17.5	20	25	30	40	50	60	70	80	90	100
7,500	.18	.15	.13	.11	.10									
10,000	.31	.26	.22	.19	.17	.14	.12	.10						
12,500	.48	.40	.34	.30	.26	.22	.18	.16	.14	.12	.11	.10		
15,000	.68	.57	.49	.43	.38	.31	.26	.23	.20	.18	.16	.15	.14	.12
17,500	.92	.77	.66	.58	.51	.42	.35	.31	.27	.24	.22	.20	.18	.16
20,000	1.20	1.00	.85	.75	.66	.54	.46	.40	.35	.31	.28	.26	.24	.21
25,000		1.57	1.34	1.17	1.04	.85	.72	.62	.55	.49	.45	.41	.37	.33
30,000			1.89	1.65	1.47	1.20	1.01	.88	.77	.69	.63	.57	.53	.47
35,000			2.56	2.24	1.98	1.62	1.37	1.19	1.05	.94	.85	.77	.71	.64
40,000				2.91	2.59	2.12	1.79	1.55	1.37	1.22	1.11	1.01	.93	.83
50,000					4.00	3.27	2.78	2.39	2.11	1.89	1.71	1.56	1.44	1.28
60,000						4.69	3.96	3.43	3.03	2.71	2.45	2.24	2.06	1.84
80,000								6.10	5.37	4.80	4.35	3.97	3.66	3.26
100,000									8.40	7.51	6.81	6.21	5.71	5.10
125,000											10.7	9.70	8.95	7.98

h. Nominal pipe size 14 in, Schedule 30

Free air, ft^3/min	Line pressure, psig													
	10	12.5	15	17.5	20	25	30	40	50	60	70	80	90	100
10,000	.18	.16	.15	.14	.13	.11	.10							
15,000	.40	.36	.33	.30	.28	.25	.22	.18	.15	.13	.12	.10		
20,000	.69	.63	.58	.53	.49	.43	.38	.31	.26	.23	.20	.18	.16	.15
25,000	1.06	.97	.89	.82	.76	.66	.59	.48	.41	.35	.31	.28	.25	.23
30,000	1.53	1.39	1.27	1.17	1.09	.95	.84	.69	.58	.51	.45	.40	.36	.33
40,000				2.06	1.91	1.67	1.48	1.21	1.02	.89	.78	.70	.64	.58
50,000					2.96	2.58	2.30	1.88	1.59	1.38	1.21	1.08	.98	.90
60,000						3.70	3.29	2.69	2.27	1.97	1.73	1.55	1.40	1.28
80,000								4.76	4.02	3.49	3.08	2.76	2.50	2.28
100,000									6.29	5.45	4.80	4.30	3.89	3.55
125,000											7.51	6.73	6.10	5.56
150,000												9.68	8.86	8.00

SOURCE: C. W. Gibbs, "New Compressed Air and Gas Data," Ingersoll-Rand Co., Phillipsburg, N.J., 1971, p. 34-82 to 34-85.

TABLE 4.3 Length of Straight Pipe in Feet Having the Same Pressure Loss as the Tabulated Fitting

Nominal pipe size, in	Schedule number	Inside diameter, in	Inside diameter, ft	Globe valve* L/D = 340	Angle valve* L/D = 145	Gate valve* L/D = 13	Swing check* valve† L/D = 135	Plug cock* L/D = 18	45° standard elbow L/D = 16	90° standard elbow L/D = 30	90° long radius elbow L/D = 20	Standard tee: Run of tee L/D = 20	Standard tee: Side outlet L/D = 60	Close return bend L/D = 50	90° welding elbow: Short radius	90° welding elbow: Long radius
½	40	0.622	0.0518	17.6	7.5	0.67	7.0	0.93	0.83	1.55	1.04	1.04	3.11	2.59		
¾	40	0.824	0.0685	23.3	9.9	0.89	9.2	1.23	1.10	2.06	1.37	1.37	4.11	3.43		
1	40	1.049	0.0872	29.7	13.6	1.14	11.8	1.57	1.40	2.62	1.74	1.74	5.2	4.36	1.4	1.1
1½	40	1.610	0.134	45.5	19.4	1.74	18.1	2.41	2.14	4.02	2.68	2.68	8.1	6.7	2.1	1.6
2	40	2.067	0.172	59	25.0	2.24	23.2	3.10	2.75	5.2	3.44	3.44	10.3	8.6	2.8	2.1
2½	40	2.469	0.206	70	29.9	2.68	27.8	3.70	3.30	6.2	4.12	4.12	12.4	10.3	3.3	2.5
3	40	3.068	0.256	87	37.1	3.32	34.6	4.60	4.10	7.7	5.1	5.1	15.4	12.8	4.1	3.1
4	40	4.026	0.335	114	48.5	4.35	45.2	6.0	5.4	10.1	6.7	6.7	20.1	16.8	5.4	4.0
5	40	5.047	0.420	143	61	5.5	57	7.6	6.7	12.6	8.4	8.4	25.2	21.0	6.7	5.1
6	40	6.065	0.505	172	73	6.6	68	9.1	8.1	15.1	10.1	10.1	30.3	25.3	8.1	6.1
8	40	7.981	0.665	226	96	8.7	90	12.0	10.7	19.9	13.3	13.3	40.0	33.3	11	8.0
10	40	10.020	0.836	284	121	10.9	113	15.0	13.4	25.1	16.7	16.7	50.2	41.8	13	10
12	40	11.938	0.995			13.0	134	17.9	15.9	29.8	19.9	19.9	60	50	16	12
14	30	13.250	1.104			14.3	149		17.7	33.2	22.1	22.1	66	55	18	13
16	30	15.250	1.270			16.5	171		20.3	38.2	25.4	25.4	76	64	20	15
18	30	17.124	1.430			18.6	193		22.8	43.2	28.6	28.6	86	72	23	17
20	20	19.250	1.600			20.8	216		25.6	48.0	32.0	32.0	96	80	25	19
24	20	23.250	1.940			25.2	262		31.0	58	38.8	38.8	117	97	30	23

* All valves and cocks to be fully open.
† Check valves require 0.50 lb/in^2 pressure loss to open fully. Welding elbow data from Midwest Piping Catalog 61 (1961). *L/D* values from Crane Co. Technical Paper No. 410 (1957). Both *L* and *D* in ft.
SOURCE: C. W. Gibbs, "New Compressed Air and Gas Data," Ingersoll-Rand Co., Phillipsburg, N.J., 1971, p. 34–78.

TABLE 4.4 Pressure Loss in Hose, Lubrication Only at Tool (No Line Lubricator)

Free air, ft^3/in	Line pressure, psig						
	60	80	100	120	150	200	300
			Hose length 50 ft, inside diameter ¾ in				
60	3.1	2.4	2.0				
80	5.3	4.2	3.5	2.9	2.4	1.8	1.2
100	8.1	6.4	5.2	4.5	3.6	2.8	1.9
120		9.0	7.4	6.3	5.1	3.9	2.7
140		12.0	9.9	8.4	6.9	5.3	3.6
160			12.7	10.8	8.9	6.8	4.6
180				13.6	11.1	8.5	5.8
200				16.6	13.5	10.4	7.1
220					16.2	12.4	8.4
			Hose length 50 ft, inside diameter 1 in				
120	2.7	2.1					
150	4.1	3.2	2.7	2.3			
180	5.8	4.6	3.8	3.2	2.6	2.0	1.3
210	7.7	6.1	5.0	4.3	3.5	217	1.8
240		7.9	6.5	5.5	4.5	3.4	2.3
270		9.8	8.1	6.9	5.6	4.3	2.9
300		12.0	9.9	8.4	6.9	5.3	3.6
330			11.8	10.0	8.2	6.3	4.3
360			13.9	11.9	9.7	7.4	5.0
390				13.8	11.3	8.7	5.9
420				15.9	13.0	10.0	6.8
450					14.8	11.4	7.7
			Hose length 50 ft, inside diameter 1¼ in				
200	2.4						
250	3.7	2.9	2.4	2.0			
300	5.2	4.1	3.4	2.9	2.3	1.8	1.2
350	7.0	5.5	4.5	3.8	3.1	2.4	1.6
400	8.9	7.0	5.8	4.9	4.0	3.1	2.1
450		8.8	7.3	6.2	5.0	3.9	2.6
500		10.8	8.9	7.6	6.2	4.7	3.2
550			10.7	9.1	7.4	5.7	3.9
600			12.6	10.7	8.7	6.7	4.6
650			14.6	12.4	10.2	7.8	5.3
700				14.3	11.7	9.0	6.1
750					13.3	10.2	6.9
800					15.0	11.5	7.8
			Hose length 50 ft, inside diameter 1½ in				
300	2.1						
400	3.7	2.9	2.4	2.0			

TABLE 4.4 Pressure Loss in Hose, Lubrication Only at Tool (No Line Lubricator) (*Continued*)

Free air, ft³/in	Line pressure, psig 60	80	100	120	150	200	300
500	5.6	4.4	3.7	3.1	2.5	1.9	1.3
600	8.0	6.3	5.2	4.4	3.6	2.8	1.9
700		8.5	7.0	5.9	4.9	3.7	2.5
800		10.9	9.0	7.7	6.3	4.8	3.2
900			11.2	9.5	7.8	6.0	4.1
1,000			13.6	11.6	9.5	7.3	4.9
1,100				14.0	11.4	8.8	6.0
1,200					13.6	10.4	7.1
1,300					15.8	12.1	8.3
Hose length 50 ft, inside diameter 2 in							
600	1.9						
800	3.2	2.5	2.1				
1,000	5.0	3.9	3.2	2.7	2.2	1.7	1.1
1,200	7.0	5.5	4.5	3.8	3.1	2.4	1.6
1,400	9.3	7.4	6.1	5.2	4.2	3.2	2.2
1,600		9.6	7.9	6.7	5.5	4.2	2.8
1,800		12.1	9.9	8.4	6.9	5.3	3.6
2,000			12.2	10.4	8.5	6.5	4.4
2,200			14.6	12.5	10.2	7.8	5.3
2,400				14.7	12.0	9.2	6.3
2,600					14.1	10.8	7.3
2,800					16.2	12.4	8.5
Hose length 50 ft, inside diameter 2½ in							
1,000	1.7						
1,500	3.7	2.9	2.4	2.0			
2,000	6.5	5.1	4.2	3.6	2.9	2.2	1.5
2,500	10.0	7.9	6.5	5.5	4.5	3.4	2.3
3,000		11.2	9.3	7.9	6.4	4.9	3.3
3,500			12.4	10.6	8.7	6.6	4.5
4,000				13.7	11.2	8.6	5.8
4,500					14.0	10.7	7.3
Hose length 50 ft, inside diameter 3 in							
2,000	2.5	2.0					
2,500	3.9	3.0	2.5	2.1			
3,000	5.5	4.4	3.6	3.1	2.5	1.9	1.3
3,500	7.5	5.9	4.9	4.1	3.4	2.6	1.7
4,000	9.8	7.6	6.3	5.3	4.4	3.3	2.3
4,500		9.6	7.9	6.7	5.5	4.2	2.8
5,000		11.7	9.6	8.2	6.7	5.1	3.5
5,500			11.5	9.8	8.0	6.1	4.2
6,000			13.6	11.5	9.4	7.2	4.9

TABLE 4.4 Pressure Loss in Hose, Lubrication Only at Tool (No Line Lubricator) (*Continued*)

Free air, ft³/in	Line pressure, psig						
	60	80	100	120	150	200	300
6,500				13.5	11.0	8.4	5.7
7,000				15.6	12.7	9.8	6.6
7,500					14.5	11.1	7.6
Hose length 25 ft, inside diameter 4 in							
5,000	1.9						
6,000	2.7	2.1	1.7				
7,000	3.6	2.8	2.3	2.0		1.2	
8,000	4.7	3.7	3.0	2.6	2.1	1.6	
9,000	5.9	4.6	3.8	3.2	2.6	2.0	
10,000	7.2	5.7	4.7	4.0	3.2	2.5	
11,000	8.7	6.8	5.6	4.8	3.9	3.0	
12,000		8.1	6.7	5.7	4.6	3.5	
13,000		9.4	7.8	6.6	5.4	4.1	
14,000			9.0	7.6	6.2	4.8	
15,000				8.7	7.1	5.4	
16,000				9.8	8.0	6.2	
17,000					9.1	6.9	

SOURCE: C. W. Gibbs, "New Compressed Air and Gas Data," Ingersoll-Rand Co., Phillipsburg, N.J., 1971, p. 34-162 to 34-163.

TABLE 4.5 Pressure Loss in Hose with Air-Line Lubricator (Oil Fog in Air)

Free air, ft³/min	Line pressure, psig						
	60	80	100	120	150	200	300
Hose length 50 ft, inside diameter ¾ in							
50	2.7	2.1					
60	4.1	3.2	2.6	2.2			
70	5.9	4.7	3.8	3.3	2.7	2.0	1.4
80	8.2	6.4	5.3	4.5	3.7	2.8	1.9
90		8.6	7.1	6.0	4.9	3.8	2.6
100		11.2	9.2	7.8	6.4	4.9	3.3
110			11.7	9.9	8.1	6.2	4.2
120			14.5	12.3	10.1	7.7	5.2
130				15.1	12.3	9.4	6.4
140					14.8	11.4	7.7
Hose length 50 ft, inside diameter 1 in							
80	1.7						
100	2.9	2.3	1.9				

TABLE 4.5 Pressure Loss in Hose with Air-Line Lubricator (Oil Fog in Air) (*Continued*)

Free air, ft³/min	Line pressure, psig						
	60	80	100	120	150	200	300
120	4.5	3.5	2.9	2.5	2.0	1.5	1.0
140	6.6	5.2	4.3	3.6	3.0	2.3	1.5
160	9.2	7.3	6.0	5.1	4.2	3.2	2.1
180		9.8	8.1	6.9	5.6	4.3	2.9
200		12.6	10.6	9.0	7.3	5.6	3.8
220			13.4	11.4	9.3	7.1	4.9
240				14.2	11.6	8.9	6.1
260					14.3	10.9	7.4
Hose length 50 ft, inside diameter 1¼ in							
150	2.3						
180	3.6	2.9	2.4	2.0			
210	5.4	4.2	3.5	3.0	2.4	1.8	1.2
240	7.5	5.9	4.9	4.2	3.4	2.6	1.7
270		8.0	6.6	5.6	4.6	3.5	2.4
300		10.4	8.6	7.3	6.0	4.6	3.1
330			11.0	9.3	7.6	5.8	4.0
360			13.7	11.7	9.6	7.3	5.0
390				14.4	11.8	9.0	6.1
420					14.3	10.9	7.5
Hose length 50 ft, inside diameter 1½ in							
220	2.2						
260	3.4	2.6	2.2				
300	4.8	3.8	3.1	2.7	2.2	1.7	1.1
340	6.6	5.2	4.3	3.7	3.0	2.3	1.5
380	8.8	7.0	5.7	4.9	4.0	3.0	2.1
420		9.0	7.4	6.3	5.1	3.9	2.7
460		11.4	9.4	8.0	6.5	5.0	3.4
500			11.7	9.9	8.1	6.2	4.2
540			14.2	12.1	9.9	7.6	5.2
580				14.6	11.9	9.1	6.2
620					14.1	10.8	7.4
Hose length 50 ft, inside diameter 2 in							
400	2.0						
480	3.2	2.5	2.0				
560	4.7	3.7	3.1	2.6	2.1	1.6	1.1
640	6.7	5.3	4.3	3.7	3.0	2.3	1.6
720	9.1	7.2	5.9	5.0	4.1	3.1	2.1
800		9.4	7.8	6.6	5.4	4.1	2.8
880			10.0	8.5	6.9	5.3	3.6
960			12.6	10.7	8.7	6.7	4.5

TABLE 4.5 Pressure Loss in Hose with Air-Line Lubricator (Oil Fog in Air) (*Continued*)

Free air, ft³/min	Line pressure, psig 60	80	100	120	150	200	300
1040				13.2	10.8	8.3	5.6
1120					13.2	10.1	6.9
Hose length 50 ft, inside diameter 2½ in							
700	2.4						
800	3.5	2.7	2.2				
900	4.7	3.7	3.0	2.6	2.1	1.6	1.1
1000	6.2	4.9	4.0	3.4	2.8	2.1	1.4
1100	7.9	6.2	5.1	4.4	3.6	2.7	1.8
1200		7.8	6.5	5.5	4.5	3.4	2.3
1300		9.7	8.0	6.8	5.6	4.3	2.9
1400		11.8	9.8	8.3	6.8	5.2	3.5
1500			11.7	10.0	8.2	6.2	4.2
1600			14.0	11.9	9.7	7.4	5.1
1700				14.0	11.4	8.8	6.0
1800					13.3	10.2	6.9
Hose length 50 ft, inside diameter 3 in							
1000	2.2						
1200	3.5	2.8	2.3				
1400	5.3	4.2	3.4	2.9	2.4	1.8	1.2
1600	7.5	5.9	4.9	4.2	3.4	2.6	1.8
1800	10.5	8.2	6.7	5.7	4.7	3.6	2.4
2000		10.8	8.9	7.6	6.2	4.8	3.2
2200			11.5	9.8	8.0	6.1	4.1
2400			14.4	12.3	10.0	7.7	5.2
2600				15.2	12.4	9.5	6.5
2800					15.1	11.6	7.9
Hose length 25 ft, inside diameter 4 in							
2400	2.2						
2800	3.3	2.6	2.1				
3200	4.7	3.7	3.0	2.6	2.1	1.6	1.1
3600	6.4	5.0	4.2	3.5	2.9	2.2	1.5
4000	8.5	6.7	5.5	4.7	3.8	2.9	2.0
4400		8.7	7.1	6.1	5.0	3.8	2.6
4800			9.0	7.7	6.3	4.8	3.3
5200				9.6	7.8	6.0	4.1
5600					9.6	7.3	5.0

SOURCE: C. W. Gibbs, "New Compressed Air and Gas Data, Ingersoll-Rand Co., Phillipsburg, N.J., 1971 p. 34-164 to 34-165.

Before using the table to determine the pressure loss in the 150 ft (45.7 m) of 2½-in (63-mm) bid hose we must divide the total cubic feet per minute available at the header by the number of drills.

$$\frac{4800 \text{ total}}{5 \text{ drills}} = 960 \text{ ft}^3\text{/min per drill}$$

Turning to Table 4.5 we see that for a 2½-in (63-mm) hose with an air-line lubricator, there is a value of approximately 3.45. Therefore, 107 − 3.45 = 103.55 for 960 ft^3/min at 107 psig. Therefore, the air received by each of the drills is 960 ft^3/min at 103.55 psig. This exceeds the required air, and therefore there should be maximum production provided by each drill.

The problem that could have arisen is the problem of not enough air. The alternatives in that case are not just to reduce the number of drills or to increase the number or size of the compressors. When short of air, before making any capital equipment changes, investigate the practicality of increasing the size of the air pipeline used. As you are probably aware, the smaller the diameter of the pipe, the greater the friction loss, and conversely, the larger the pipe diameter, the less the friction loss.

For example, if we had chosen to use 4-in (102-mm) pipe instead of 6-in (152-mm) we would not have had enough air. Given 4800 ft^3/min at 110 psig in 4-in pipe, interpolating we obtain:

$$n \times p = P$$

$$7 \times 3.88 = 27.16$$

$$110 - 27.16 = 82.82 \text{ psig}$$

not enough pressure to efficiently run the drills.

When selecting the compressor to fill the air requirement, first one must take into account, not only the total cubic feet per minute required, but also the costs. In our example, we used four 1200 ft^3/min compressors, which are probably the most readily available and less costly in combination than other units. To use all 750 ft^3/min compressors would require at least six units, which might necessitate another compressor operator. (See Chapter 13.) To use large compressors, say three 1800 ft^3/min, might be impractical because of their lesser availability and the sheer logistics of moving the compressors. Twelve hundred ft^3/min is generally the largest portable compressor that can be moved relatively easily.

In conclusion, selecting a compressor is not just matching air requirements. It also includes economics, logistics, and practicality.

FIVE

FIRING SYSTEMS

BLASTING CAPS

Blasting caps are small cylindrical tubes that detonate cap-sensitive explosives. They are usually made of copper or aluminum. The blasting cap contains an explosive, which when ignited will produce an explosion with enough intensity of heat and shock to detonate the cap-sensitive explosive. There are three types of blasting caps in use today: common caps, millisecond delays (MS delays), and standard delays.

Common Blasting Caps

The common blasting cap is detonated by a fuse. It is the earliest of the modern caps, and therefore now the least common. The common cap is simply a copper or aluminum cylinder approximately 1½ in (38 mm) long by ¼ in (6 mm) in diameter and closed at one end. The cap contains two types of charges: the igniting charge and the base charge. A safety fuse is inserted into the open end of the cap. When the fuse is ignited, the powder core burns, acting as the vehicle through which the fire is transmitted to the igniting charge end of the cap. The burning fuse spits a flame resembling a jet flame, called an "ignition spit." (See Figure 5.1.) When the flame travels to the cap, it ignites the ignition charge, which detonates the base charge, which in turn detonates the explosive charge that is being primed with the cap.

The safety fuse (which was invented by William Bickford to ignite black powder) is inserted into a common cap with the end cut squarely to ensure that the flame reaches the igniting charge completely. (See Figures 5.2 and 5.3.) The fuse, once cut properly, is inserted into the cap carefully, so that it butts against the charge in the cap. Once the fuse is properly seated the cap is crimped with a crimping tool (see Figures 5.4 and 5.5), to prevent the fuse from slipping out of the cap and to prevent water or other foreign material from entering the cap and inhibiting detonation.

The safest way to determine the burning time of a safety fuse is to cut pieces of fuse from the coil intended for use, and to light and time them.

Fig. 5.1 Fuse flame jet. The flame jets and ignites the ignition charge.

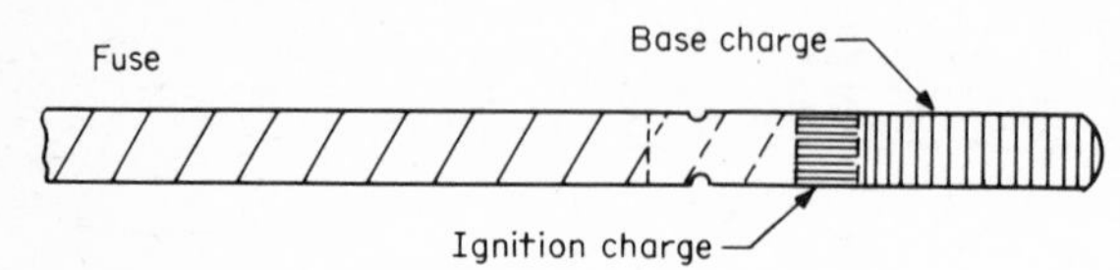

Fig. 5.2 Cap with safety fuse.

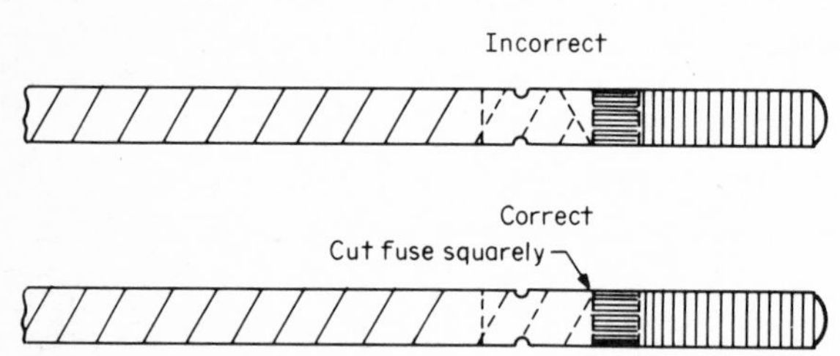

Fig. 5.3 Proper placing of safety fuse in cap.

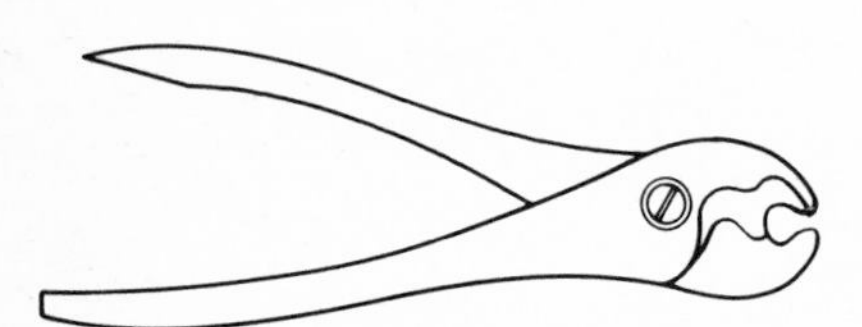

Fig. 5.4 Crimping tool.

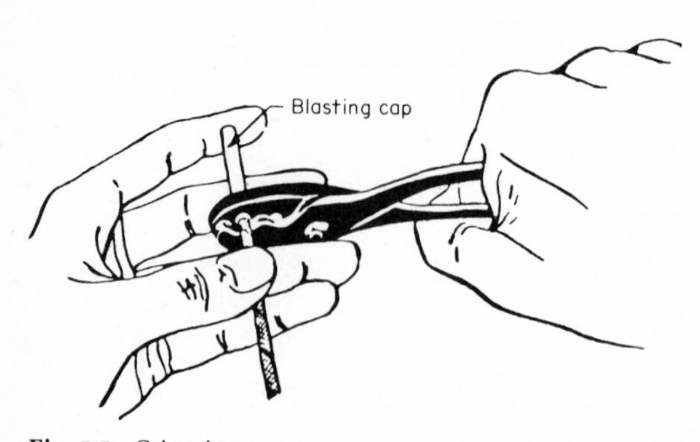

Fig. 5.5 Crimping a cap.

Burning times can vary, and one should never trust the safety fuse to burn to the time it is specified to burn.

The common cap, with safety fuse, offers the blaster the opportunity to light the fuse and reach safety, assuming enough time was allowed. Today common caps are rarely used, except in agriculture or in cases where firing electrically is prohibited (sometimes when blasting on a military base, for example, the contractor may be restricted from using electric blasting caps because of the danger of induced currents caused by the radio transmitters).

Electric Blasting Caps

The need for a more controllable method of blast detonation gave rise to the development of electric blasting caps. The electric cap can be called a cross between a common cap and a light bulb. The cap contains charges like the common cap, but instead of a safety fuse the cap contains two wires that meet at a bridge wire. When electric current is applied, the bridge wire burns much as a light bulb, igniting the charge in the cap. This method of detonation enables the blaster to detonate the blast when it is most advantageous. Instead of lighting a fuse and trying to obtain cover before the fuse burns down and ignites the cap, the blaster may move to safety and apply the electric current at any time desired. The electric blasting cap also enables the blaster to shoot more holes than the safety fuse method.

Delay Blasting Caps

Delay blasting caps are caps that are detonated by electricity in various time-delay intervals. The leg wires that extend out from the cap are usually copper and are used to transmit the electricity to the cap from the power source. The bridge wire on a delay cap is coated with a substance that requires a calculated time to burn off before the bridge wire can be exposed to ignite the charge in the cap. By using various amounts of coating and various lengths of bridge wire you can set caps to ignite at different time intervals. These intervals are called "delays." There are two types of delays: standard and millisecond. The standard delay is the forerunner of the millisecond delay and is usually confined to underground work. This delay is usually set so that caps detonate at approximately ½-s intervals.

The millisecond delay is calculated so that caps detonate at intervals ranging from 8 to 50 milliseconds (ms) apart, depending upon the manufacturer. The use of millisecond-delay blasting caps offers various advantages over the common standard-delay blasting caps:

1. They reduce ground vibration.
2. They improve fragmentation.
3. They produce less flyrock.
4. They reduce costs.
5. They reduce overbreak.

TABLE 5.1 Typical Delay Intervals with Short Period Delay Detonators

Delay no.	Atlas (Rockmaster) Interval, msec	Atlas (Rockmaster) Elapsed time, msec	C-I-L (short period caps no. 8) Interval, msec	C-I-L (short period caps no. 8) Elapsed time, msec	DuPont (MS delay) Interval, msec	DuPont (MS delay) Elapsed time, msec	Hercules (Millidet no. 8) Interval, msec	Hercules (Millidet no. 8) Elapsed time, msec
0	—	0	—	0	—	0	—	0
1	8	8	8	8	25	25	12	12
2	17	25	22	30	25	50	13	25
3	25	50	20	50	25	75	25	50
4	25	75	25	75	25	100	25	75
5	25	100	25	100	25	125	25	100
6	25	125	30	130	25	150	35	135
7	25	150	30	160	25	175	35	170
8	25	175	30	190	25	200	35	205
9	25	200	40	230	50	250	35	240
10	50	250	50	280	50	300	40	280
11	50	300	60	340	50	350	40	320
12	50	350	70	410	50	400	40	360
13	50	400	80	490	50	450	40	400
14	50	450	80	570	50	500	50	450
15	50	500	80	650	100	600	50	500
16	50	550	75	725	100	700	50	550
17	100	650	75	800	100	800	50	600
18	100	750	75	875	100	900	100	700
19	125	875	75	950	100	1000	—	—
20	125	1000	75	1025	—	—	—	—
21	125	1125	100	1125	—	—	—	—
22	125	1250	100	1225	—	—	—	—
23	125	1375	125	1350	—	—	—	—
24	125	1500	150	1500	—	—	—	—
25	125	1625	175	1675	—	—	—	—
26	125	1750	200	1875	—	—	—	—
27	125	1875	200	2075	—	—	—	—
28	125	2000	225	2300	—	—	—	—
29	125	2125	250	2550	—	—	—	—
30	125	2250	330	2880	—	—	—	—
31	125	2375	230	3050	—	—	—	—
to	125		—	—	—	—	—	—
38	125	3250	—	—	—	—	—	—

SOURCE: C. E. Gregory, "Explosives for North American Engineers," 2d ed., Trans Tech Publications, Clausthal, West Germany, 1973, p. 107.

Table 5.1 contains the different delay intervals for various common commercial electric blasting caps.

Delay Blasting

The amount of vibration caused by blasting is related to the amount of explosives detonated at any one time. Through the use of millisecond delays the amount of explosives fired at one given instant may be reduced by using a different delay period for different holes. The standard today requires 9 ms between blastholes for each to be considered an independent blast with regard to vibration. For example, if a blast is to be composed of four holes, each containing 25 lb (11 kg) of explosive, when they are detonated simultaneously a total of 100 lb (45 kg) of explosive is detonated at one time. However, if each hole is loaded with the same amount of explosive but has a different delay from the others, then there is only 25 lb (11 kg) of explosives being detonated at one instant. Thus the use of delays permits a larger total blast, because the amount of explosive being detonated at one time is reduced.

By permitting the movement of rock at various time intervals the millisecond delay cap aids fragmentation. As will be discussed in Chapter 6, the rock must be free to bend into a flexural rupture to maximize fragmentation. If a blaster wishes to shoot more than one row of holes at a time, as is usually the case, it is necessary to have a delay period to permit the rock of each row to move without restriction. In Figure 5.6, without a delay in the firing between the two rows, the rock of the second row cannot move toward the face. In this case this would usually result in flyrock from the top of the cut, and the remainder would be poorly fragmented. In Figure 5.7, where delays are used, the second row is permitted to move into a state of flexural rupture because the first has already ruptured and moved away from the rock formation.

Researchers believe that the delay interval necessary for optimum frag-

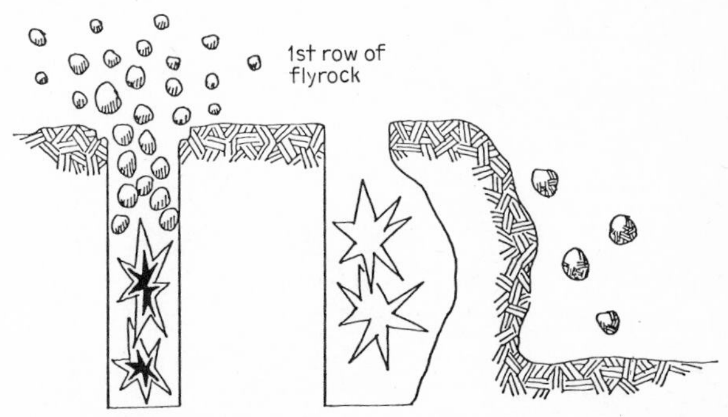

Fig. 5.6 Flyrock caused by not using delay blasting techniques. Second row cannot move toward face and therefore must either fly or remain in place.

mentation varies with the type of rock and burden distances. However, it is thought that the optimum delay interval lies between 1 and 2 ms per foot of burden. Therefore, with the use of proper delay intervals, fragmentation can be greatly improved.

Flyrock is the excessive movement of blasted rock. In blasting, rock is expected to move to create breakage; however, when rock moves excessively, generally in the air, it is called "flyrock." The use of delays can help reduce flyrock by permitting the rock to move in the direction desired rather than moving haphazardly through the air.

Referring to Figures 5.6 and 5.7, you can see how the blast without delays tends to have more flyrock. This is because the explosives detonate pushing against the rock. The rock in the second row cannot push toward the free face because the rock of the first row is still in place. Therefore, the rock, as with any force, follows the path of least resistance, which is up, resulting in undesired flyrock. In the example with delays the rock from the second row is able to move toward the free face because its detonating is delayed, permitting the first row to move away and to create a new free face for the second row. The result of the new free face is the reduction of flyrock.

Millisecond-delay blasting can reduce blasting costs substantially. By using good blasting design, you can extend burden and spacings and reduce the amount of drilling necessary. Also, delay blasting permits the firing of several holes, therefore reducing the amount of secondary work (boulders). By permitting each hole to experience flexural rupture, delay blasting reduces the amount of explosive necessary. Blasting without delays requires considerably more drilling and explosive to break the rock because the rock tends to resist breakage because of the lack of a sufficient number of free faces. Therefore, this increased powder factor would make conditions conducive for greater flyrock.

Delay blasting can reduce overbreak because it permits each hole to undergo flexural rupture. If delays were not used (as in Figure 5.6), the energy that created excess throw (flyrock) would cause the rock to fracture behind the blast. This additional undesirable breakage is called "overbreak." Generally, this overbreak is in the form of boulders, thus creating an additional expense.

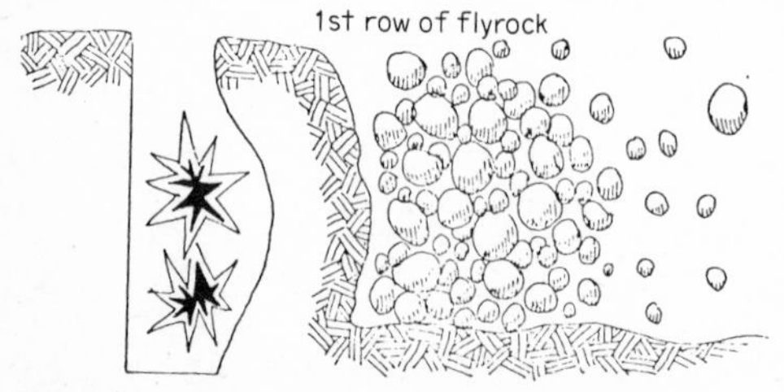

Fig. 5.7 Using delay techniques reduces flyrock.

As has been demonstrated, delay blasting plays a major role in modern blasting operations. With proper blast design, delay caps can effectively reduce costs and vibrations while increasing desired fragmentation.

DETONATING SYSTEMS

Detonating cord is a cord consisting of an inner cord of explosive, usually PETN, surrounded by a plastic encasement and wrapped in a braid of various textiles and waterproofing substances. Detonating cord is used as a medium by which to transmit explosive detonations from a blasting cap to the explosive to be detonated.

Detonating cord is much safer to handle than a blasting cap because it is less sensitive to heat or concussion. Detonating cord is a high explosive; it will burn at a speed in excess of 4 miles per second (mi/s) (21,000 to 25,000 ft/s, or 6400 to 7600 m/s).

The main components of an electric detonating system are the electric blasting cap, the blasting wire, and the power source, or blasting machine. There are three parts to the blasting wire: the cap leg wires, connecting wire, and the lead wires.

The leg wires are the two wires extending from the blasting cap that are used to connect the cap into the circuit. These wires are usually two different colors to avoid confusion when connecting the circuit. Until the circuit is being connected, the two leg wires should always be shunted (tied together) to prevent stray current from accidentally detonating the blast.

Connecting wire is, just as the term implies, wire used to connect the circuit to the lead wire. Connecting wire is usually No. 20 soft copper with a vinyl insulation. The connecting wire protects the lead wire from being damaged by the blast by allowing the lead wire to be kept off the blast. The resistance of the No. 20 copper connecting wire is approximately 10.15 ohms (Ω) per 1000 ft (300 m). For cap leg wire resistance see the Appendix.

The lead wire, or shooting line, is usually a No. 14 copper wire that acts as the medium through which the electric impulse is delivered to the circuit from the blasting machine. This wire has an approximate resistance of 2.6 Ω/1000 ft.

As should be apparent by now, it is important to know the amount of resistance in a blasting circuit. This permits the blaster to make proper selection of a blasting machine and to change circuits to fit within its capabilities.

Electrical Requirements

To understand the electrical requirements of a blasting circuit it is necessary to review basic principles of electricity. A blaster is interested in three factors: voltage, amperage, and ohms. Voltage is the amount of electrical pressure in volts (V) in a conductor and corresponds to the pressure in pounds per square inch in a compressed air system. The amperage is the rate of flow of electricity through the conductor or wire measured in amperes, or amps (A), just as the rate of flow for air through an air line is measured in cubic feet per minute. The term "ohm" (Ω) is used to define or represent the amount of resistance in the conductor to the flow of electricity. This resistance is determined by the type of material that the conductor, or wire, is made of and the cross-sectional area of the conductor.

These are the factors whose relationship is measured by Ohm's law. Ohm's law simply states that if the voltage in volts is divided by the resistance in ohms of the circuit, the quotient will be the current in amperes that flows through the circuit. The blaster is interested in this law because the amperage that flows through the circuit is of paramount importance to the complete and total detonation of the blast. If the circuit does not have sufficient amperage, not all of the caps in the circuit will be detonated. The problem would be simpler if the circuit would just not fire with inadequate amperage. However, what actually happens is that the blast will probably detonate leaving various random caps undetonated. This can be very dangerous, because the movement of the rock caused by the holes that did detonate will bury the undetonated holes. The difficulty of finding which holes did detonate and which ones did not is a nightmare. Any blaster who has ever tried to find undetonated holes tends to be very conservative when calculating how big an electric circuit the blasting machine can handle.

Fitting the Blasting Machine to the Requirements

The calculation for determining the resistance in a circuit is relatively simple. There are three types of blasting circuits: the single-series circuit, the straight parallel circuit, and the parallel-series circuit.

Series Circuit A single-series circuit is a circuit where all the current flows through every electric blasting cap. (See Figure 5.8.) The single series is the most common method for blasting a small number of holes. It is generally accepted that the minimum amperage for a series circuit is 1.5 A of dc or 3.0 A of ac current flow. Many blasters will agree that a good, conservative limit for one series is 50 caps with 24-ft (7.3-m) leg wires.

The calculation in the following example for determining the resistance of a series circuit assumes 100 ft (30 m) of connecting wire and a shooting line

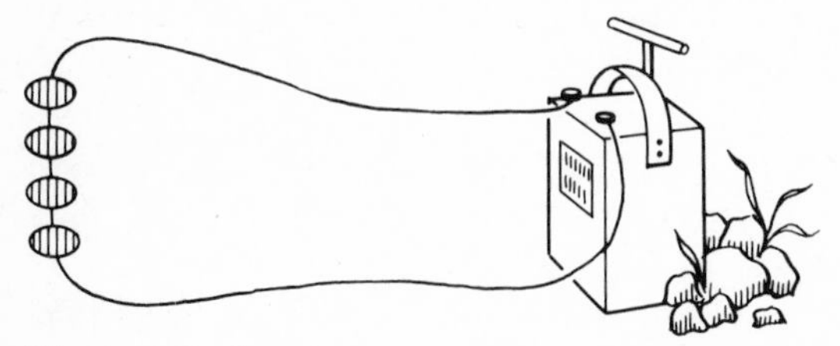

Fig. 5.8 Single series.

200 ft (60 m) long. The required voltage is to be found from the resistance calculated and the minimum amperage.

Wire	Resistance
50 caps at 2.2 Ω per cap	110 Ω
100 ft of connecting wire at 10.15 Ω/1000 ft	1.02
400 ft of lead wire at 2.5 Ω/1000 ft	1.0
Total	112.02 Ω

$$\text{Required voltage} = 1.5\text{ A (dc)} \times 112.02\ \Omega = 168.03\text{ V}$$

Alternatively, after the resistance has been calculated, if the contractor has a blasting machine of 225 V (dc), for example, then the fit of the circuit to the machine can be determined as follows:

$$\frac{225\text{ V}}{112.02\ \Omega} = 2\text{A}$$

which exceeds the minimum allowable current of 1.5 A for direct current.

Straight Parallel Circuit The parallel circuit is a circuit in which each detonator provides an alternative path for the current. This method is common in underground blasting. Two separate wires are used, with the leg wires from each cap going to each of the wires. (See Figure 5.9.) The calcula-

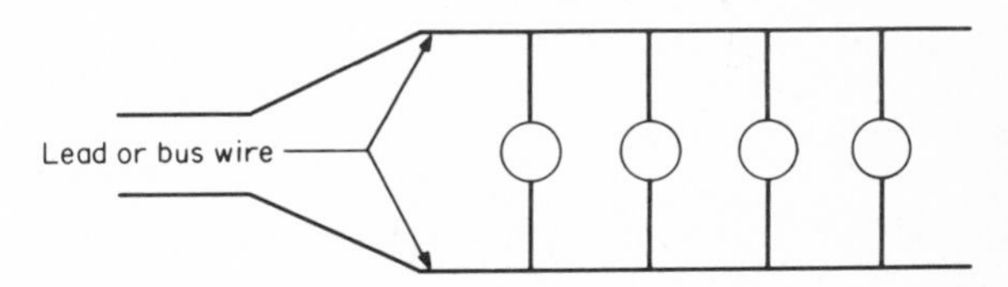

Fig. 5.9 Straight parallel circuit.

tion for this system is similar; however, the requirement differs for this series: a minimum of 1.0 A ac or dc is required for each blasting cap. For this illustration the factors will be 100 ft (30 m) of bus wire at 2.54 Ω /1000 ft, 20 caps with 12-ft (3.6-m) leg wires, 500 ft (152 m) of lead wire at 2.54 Ω /1000 ft, and 100 ft (30 m) of connecting wire at 10.15 Ω /1000 ft. When bus wire is used, its resistance is calculated by multiplying the resistance of the wire by 0.5. Again, it is to be determined whether a 225-V blasting machine fits the circuit requirements.

The resistance of the blasting caps in a straight parallel circuit is obtained by dividing the resistance of one cap by the number of caps. If in this example the resistance of one cap is 1.7 Ω, then:

$$\text{Total cap resistance} = \frac{1.7\ \Omega}{20} = 0.085\ \Omega$$

Wire	Resistance
Bus wire	0.5 (100 ft)(2.54 Ω/1000 ft) = 0.127 Ω
Caps	$\frac{1.7\ \Omega}{20} = 0.085$
Connecting wire	(100 ft)(10.15 Ω/1000 ft) = 1.1
Lead wire	2 (500 ft)(2.54 Ω/1000 ft) = 2.54
Total circuit resistance	= 3.852

To determine whether the circuit fits, first find the available amperage:

$$\frac{225\ \text{V}}{3.852\ \Omega} = 58.4\ \text{A}$$

Then divide the total amperage by the number of caps:

$$\frac{58.4\ \text{A}}{20} = 2.92\ \text{A per cap}$$

which exceeds the 1.0-A minimum. (If you use metric lengths, the results will be the same. The resistance will simply be calculated for the same length but with different units.)

Parallel-Series Circuit In a parallel-series circuit the shot is broken down into a number of equal series to permit the blasting of more holes than a single-series circuit will allow. (See Figure 5.10.) Generally the blaster will consider using the parallel-series circuit when the blast requirement exceeds 40 caps with leg wires longer than 20 ft (6 m), too many for a single-series circuit.

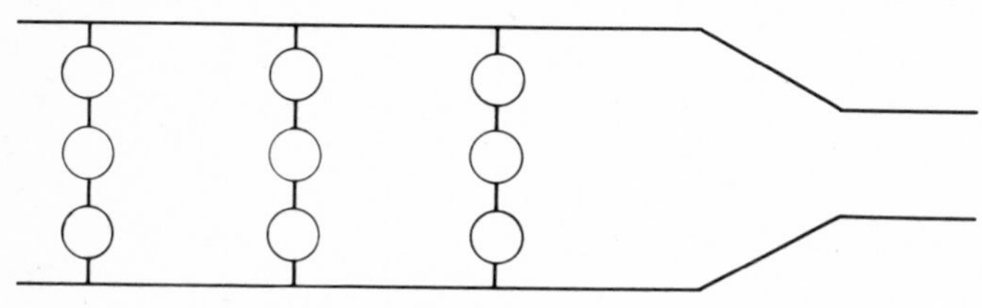

Fig. 5.10 Parallel-series circuit.

The current-flow recommendations are the same as for the single-series circuit. The resistance and current calculations for matching the parallel-series circuit to the power source are as follows:

1. Find the cap resistance in one series by multiplying the number of caps in that series by the resistance per cap.

2. Calculate the resistance of the connecting wire and shooting line as it is calculated for the single-series circuit.
3. Total the resistance of the caps, connecting wire, and lead wire.
4. Apply Ohm's law to determine the total current delivered.
5. Divide the total current delivered by the number of series to obtain the current per series.

To illustrate, assume a blast of 100 caps with 20-ft (6-m) leg wires, with a resistance of 1.9 Ω per cap; 500 ft (152 m) of connecting wire at 10.15 Ω/1000 ft; and a 200-ft (61-m) lead wire at 2.54 Ω /1000 ft. There will be four sections of 25 caps.

Calculations:

1. $25 \times 1.9\ \Omega = 47.5\ \Omega$
2. $500\ \text{ft} \times 10.15\ \Omega/1000\ \text{ft} = 5.08\ \Omega$

   $2 \times 200\ \text{ft} \times 2.54\ \Omega/1000\ \text{ft} = 1.02\ \Omega$

3. | | |
   |---|---|
   | Caps | 47.5 Ω |
   | Connecting wire | 5.08 |
   | Lead wire | 1.02 |
   | Total resistance | 53.60 Ω |

4. $$\frac{225\ \text{V (dc)}}{53.6\ \Omega} = 4.19\ \text{A}$$

5. $$\frac{4.19\ \text{A}}{4\ \text{series}} = 1.04\ \text{A per series}$$

This circuit does not fit the power source.

This method is quite effective for less than 500 holes. However, the tying in and computations become more complex for a shot that contains several thousand holes. These extremely large blasts require compensation for uneven current distribution and also the number of series for various power sources becomes limited. However, the previously mentioned calculations are adequate for most blasting operations. When blasting an extremely large number of holes, the blasting cap manufacturer should be consulted.

NONEL DETONATION

In an effort to help reduce the problems related to electric delay detonation systems, a system of nonelectric delay initiation was developed by Nitro Nobel, A.B., of Sweden. This system, called Nonel, uses a thin, transparent plastic tube that has an outside diameter of 0.12 in and an inside diameter of 0.08 in. This tube contains a thin coating, 0.1 g/ft, of reactive material, which when initiated by a blasting cap or detonating cord transmits an impulse, or shock wave, which in turn transmits a low energy signal from one point to another at a velocity of 6000 ft/s. The tube is attached to a nonelectric delay detonator, which is detonated by this shock wave. The advantages of this

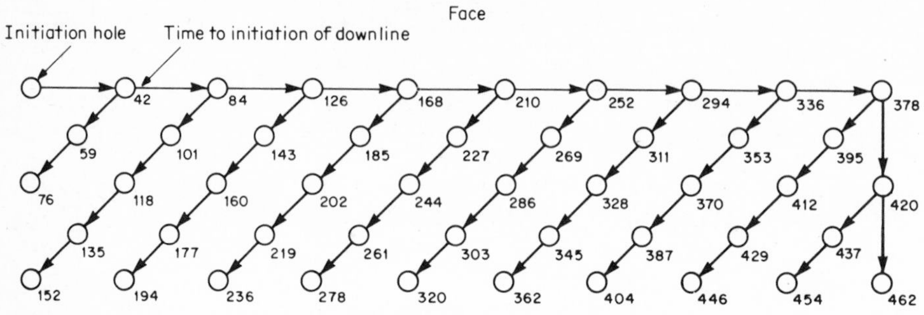

Fig. 5.11 Sequential initiation, hole by hole, five rows, offset pattern. (*Ensign Bickford Co.*)

type of system are obvious: (1) it is safe from stray electrical currents and radio-frequency hazards; (2) it is insensitive to concussion and heat whether confined or unconfined; and (3) the Nonel tube will not detonate any commercial explosives, including sensitive nitroglycerin dynamites.

The Nonel system includes various Nonel-based nonelectric delay systems for both surface and underground application. The underground systems will be discussed in Chapter 12.

The HD Nonel Primadet system is designed for use in all surface applications (strip mines, quarries, and construction). The noiseless trunk line permits the utilization of delay intervals on a trunk line, which does not have to be covered to prevent noise. When used with MS connectors, the HD Nonel Primadet system's sequence of delays can be almost infinite. This is because there is an MS-delay connector in the trunk line branches in addition to the delay in the primer itself.

Figure 5.11 shows a typical delay period layout using the same delay for the primer but varying the trunk line delays. The system can also be used

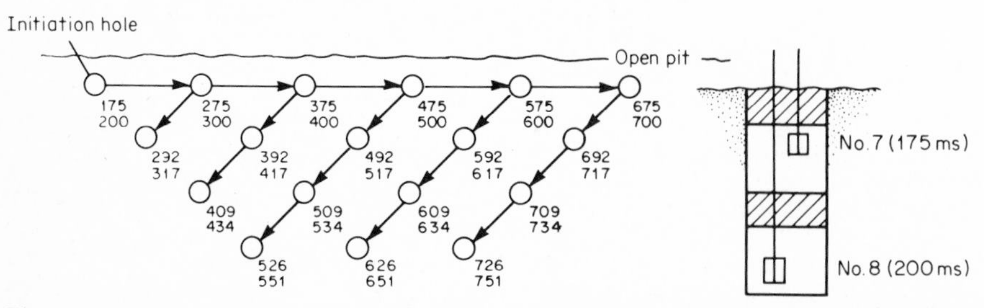

Fig. 5.12 Same delay, each hole with decks. If decked explosive charges are needed for vibration control, then the sequence of intervals shown here may be used. An unlimited sequence of intervals using up to eight delays in each hole and multiple rows may be designed whereby no two decked charges detonate simultaneously. (*Ensign Bickford Co.*)

TIME SCHEDULE*

MS time overlap zone

Master Cable position and Circuit No.:	1		2		3		4		5		6		7		8		9		10		
Circuit Time:	0		60		120		180		240		300		360		420		480		540		
	+	–	+	–	+	–	+	–	+	–	+	–	+	–	+	–	+	–	+	–	
Cap ms:	500	550	500	550	500	550	500	550	500	550	500	550	500	550	500	550	500	550	500	550	Row
Det. time:	500	550	560	610	620	670	680	730	740	790	800	850	860	910	920	970	980	1030	1040	1090	5
Cap ms:	400	450	400	450	400	450	400	450	400	450	400	450	400	450	400	450	400	450	400	450	Row
Det. time:	400	450	460	510	520	570	580	630	640	690	700	750	760	810	820	870	880	930	940	990	4
Cap ms:	300	350	300	350	300	350	300	350	300	350	300	350	300	350	300	350	300	350	300	350	Row
Det. time:	300	350	360	410	420	470	480	530	540	590	600	650	660	710	720	770	780	830	840	890	3
Cap ms:	200	250	200	250	200	250	200	250	200	250	200	250	200	250	200	250	200	250	200	250	Row
Det. time:	200	250	260	310	320	370	380	430	440	490	500	550	560	610	620	670	680	730	740	790	2
Cap ms:	100	150	100	150	100	150	100	150	100	150	100	150	100	150	100	150	100	150	100	150	Row
Det. time:	100	150	160	210	220	270	280	330	340	390	400	450	460	510	520	570	580	630	640	690	1

Open face

1st cap or hole No. 1

- - - - Open face - - - -

Cap ms—Actual time tag on cap.

Det. time—Detonation time as sequenced by the BM-125-10. Total time on this plan = 1090 ms, less starting time of 100 ms.

1000 ms = 1s

Mode of operation – Circuit No. 1 = 0 ms + cap time; Circuit No. 2 = 6 ms + cap time, etc.

* Please note on this typical layout that the timing doubles up starting with the sixth circuit, as the 400 ms in the first row matches the 400 ms in the sixth row. The blast pattern, timing, and other data are for illustration only. Success and accuracy of the blast and timing are not intended. MS- cap error and other factors must be calculated prior to all blasting operations.

Fig. 5.13 Typical delay pattern (*Research Energy of Ohio, Inc.*)

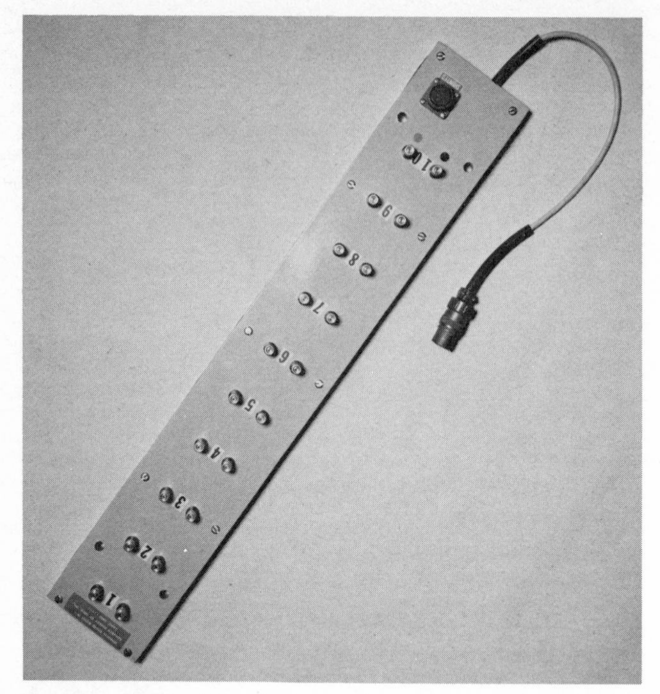

Fig. 5.14 Terminal board. (*Research Energy of Ohio, Inc.*)

with decks, as in Figure 5.12, to decrease the vibration.[1] A typical delay pattern is shown in Figure 5.13.

THE SEQUENTIAL BLASTING MACHINE

The sequential blasting machine, as developed by Research Energy of Ohio, Inc., is a condenser-discharge blasting machine with a sequential timer that permits the detonation of many electric caps at different, precisely timed intervals. The machine consists of 10 different firing circuits that are programmed to fire one after another at selected intervals. The combination of 10 different circuits, or intervals, in conjunction with delay blasting caps can yield many independent blasts. Figure 5.14 shows the terminal board of a sequential blasting machine, with its connections for 10 different circuits, and Figure 5.15 shows a complete sequential unit.

The sequential blasting machine can be set to fire at 10- to 200-ms intervals.

[1] Before attempting to use these or any explosive products the technical bulletins should be studied and understood.

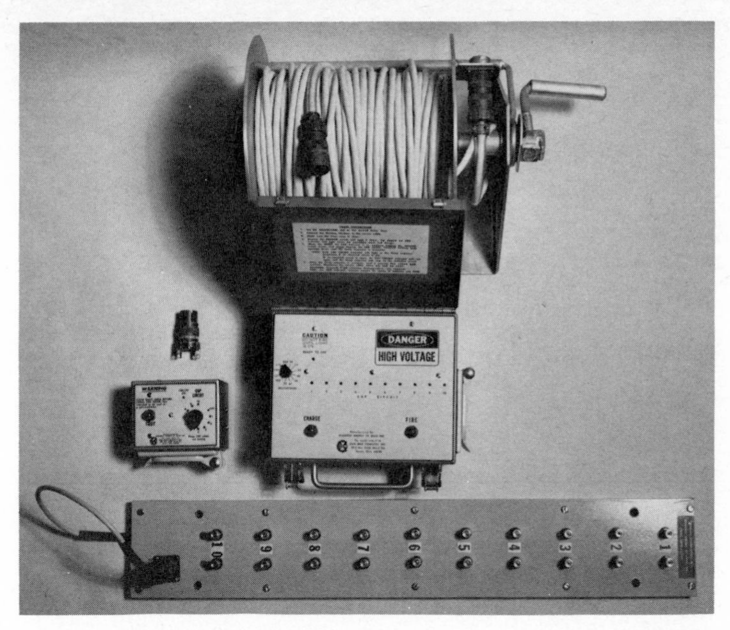

Fig. 5.15 Complete sequential blasting unit with circuit tester, cable reel, terminal board, and blasting machine. (*Research Energy of Ohio, Inc.*)

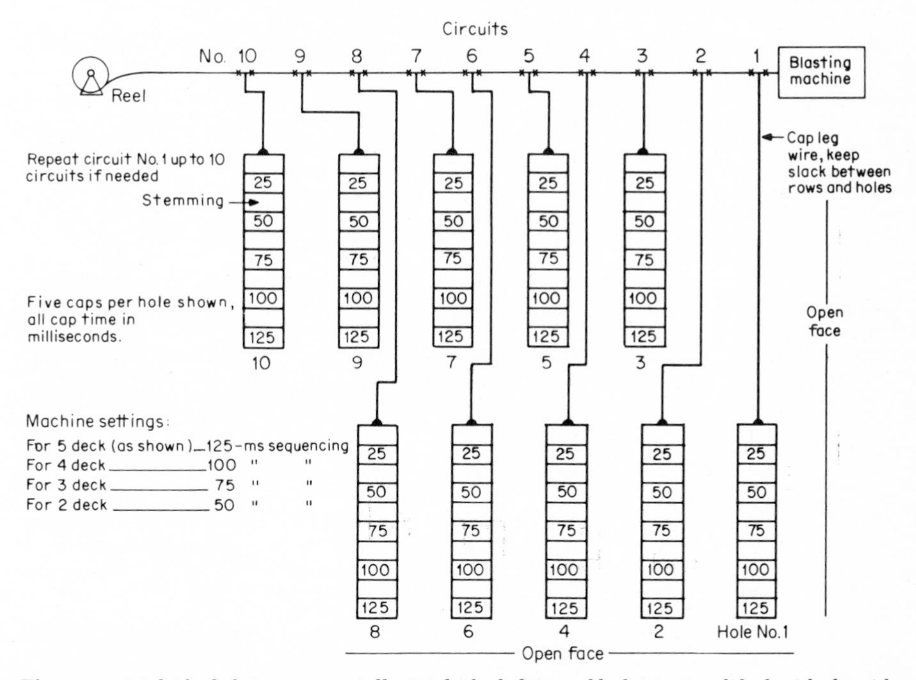

Fig. 5.16 Multideck firing sequentially. Multideck firing of holes is simplified with the aid of sequential timing. This type of blasting machine permits detonating much larger blasts per hole than normally permitted by law. With a five-deck plan, the holes are triggered 125 ms apart. Each hole also has the same series of delay caps, 25-50-75-100-125, and the caps go off from top to bottom. As the last cap in hole No. 1 goes off (at 125 ms), the caps in hole No. 2 are triggered, so that the top cap in hole No. 2 goes off at 150 ms. (*Coal Age.*)

When you use sequential timing, the same rules apply for blast design as with delay blasting caps. The sequential blasting machine makes available a great number of delay intervals, thus permitting great flexibility in blast design. Figure 5.16 is a diagram of a sequential blast setup using decking. Note how greatly the powder factor can be reduced by the utilization of both the electric delay blasting cap and the sequential blasting machine.

When you use a sequential blasting machine it is important that great care be given to the blast design and the tie-in. If the tie-in is not done by a person with a complete understanding of the wiring plan for the blast and an error is made, there may be a cutoff, resulting in only partial detonation, leaving "live" holes in the ground. The sequential machine does not transmit the impulse until the delay has passed and is therefore subject to the rock movement, which may break wires before the energy has been transmitted to the cap, whereas with electric delay blasting caps the delay is in the cap itself, preventing cutoff. Therefore, when the sequential blasting machine is used it is imperative that the shot be wired correctly and that there be some slack left in the wires to allow for rock movement. (See Figure 5.17.)

The sequential blasting machine has been used with success in blasting at open-pit mines, quarries, and construction sites, and for any application that requires a greater selection of delay intervals.

Figures 5.18 to 5.20 are examples of various timings and designs using the sequential blasting machine.

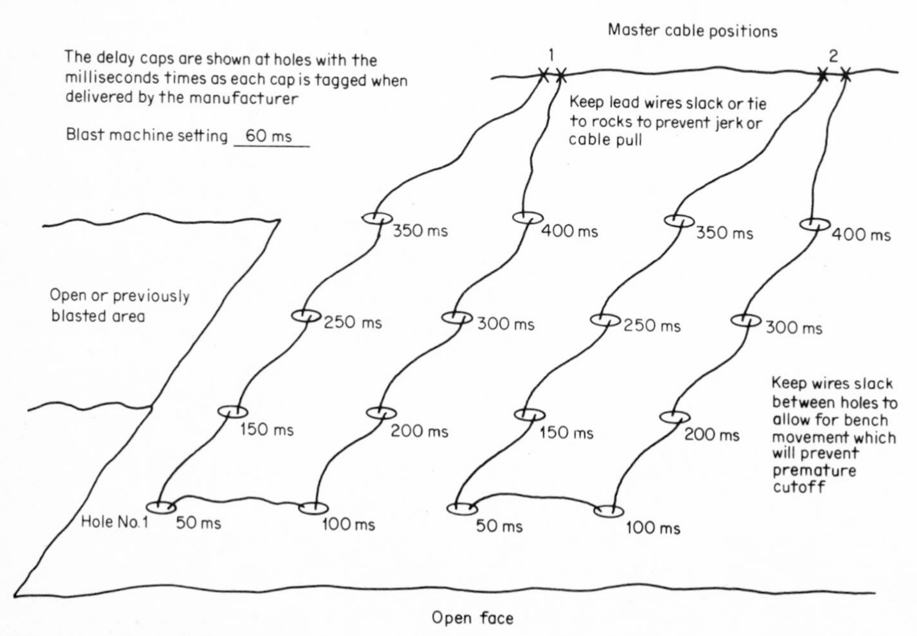

Fig. 5.17 Typical blast pattern. For critical blast, i.e., where vibration is a problem, calculate scale distance for powder loads. (*Research Energy of Ohio, Inc.*)

Circuit No. 1: 0 ms

1	6	11	16	21	26	31	36	41	46	51	56	61	66	71
●	●	●	●	●	●	●	●	●	●	●	●	●	●	●
50	100	150	200	250	300	350	400	450	500	600	700	800	900	1000

Circuit No. 2: 10 ms

2	7	12	17	22	27	32	37	42	47	52	57	62	67	72
●	●	●	●	●	●	●	●	●	●	●	●	●	●	●
60	110	160	210	260	310	360	410	460	510	610	710	810	910	1010

Circuit No. 3: 20 ms

3	8	13	18	23	28	33	38	43	48	53	58	63	68	73
●	●	●	●	●	●	●	●	●	●	●	●	●	●	●
70	120	170	220	270	320	370	420	470	520	620	720	820	920	1020

Circuit No. 4: 30 ms

4	9	14	19	24	29	34	39	44	49	54	59	64	69	74
●	●	●	●	●	●	●	●	●	●	●	●	●	●	●
80	130	180	230	280	330	380	430	480	530	630	730	830	930	1030

Circuit No. 5: 40 ms

5	10	15	20	25	30	35	40	45	50	55	60	65	70	75
●	●	●	●	●	●	●	●	●	●	●	●	●	●	●
90	140	190	240	290	340	390	440	490	540	640	740	840	940	1040

1 — Order holes
●
50 — Shooting time

Du Pont MS delay caps used

Cap No.	ms	Cap No.	ms	Cap No.	ms
2	50	10	300	15	600
4	100	11	350	16	700
6	150	12	400	17	800
8	200	13	450	18	900
9	250	14	500	19	1000

Fig. 5.18 Five rows, 75 holes, 75 delays; minimum time between delays, 10 ms. (Set blasting machine on 10 ms. All circuits energized before first hole shoots.) (*Research Energy of Ohio, Inc.*)

Circuit No. 1: 0 ms

1	4	7	10	13	16	19	22	25	28	31	34	37	40	43
●	●	●	●	●	●	●	●	●	●	●	●	●	●	●
50	100	150	200	250	300	350	400	450	500	600	700	800	900	1000

Circuit No. 2: 17 ms

2	5	8	11	14	17	20	23	26	29	32	35	38	41	44
●	●	●	●	●	●	●	●	●	●	●	●	●	●	●
67	117	167	217	267	317	367	417	467	517	617	717	817	917	1017

Circuit No. 3: 34 ms

3	6	9	12	15	18	21	24	27	30	33	36	39	42	45
●	●	●	●	●	●	●	●	●	●	●	●	●	●	●
84	134	184	234	284	334	384	434	484	534	634	734	834	934	1034

1 – Order holes shoot
●
50 – Shooting time

Du Pont MS delay caps used

Cap No.	ms	Cap No.	ms	Cap No.	ms
2	50	10	300	15	600
4	100	11	350	16	700
6	150	12	400	17	800
8	200	13	450	18	900
9	250	14	500	19	1000

Fig. 5.19 Three rows, 45 holes, 45 delays; minimum time between delays, 17 ms. (Set blasting machine on 17 ms. Two circuits energized before first hole shoots.) (*Research Energy of Ohio, Inc.*)

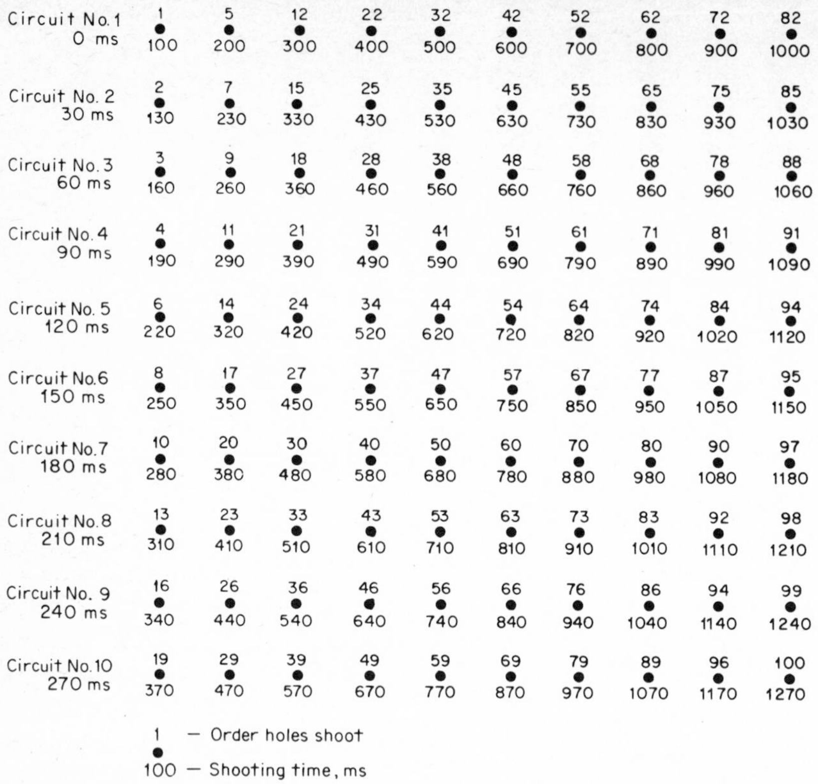

Circuit										
Circuit No. 1 0 ms	1 (100)	5 (200)	12 (300)	22 (400)	32 (500)	42 (600)	52 (700)	62 (800)	72 (900)	82 (1000)
Circuit No. 2 30 ms	2 (130)	7 (230)	15 (330)	25 (430)	35 (530)	45 (630)	55 (730)	65 (830)	75 (930)	85 (1030)
Circuit No. 3 60 ms	3 (160)	9 (260)	18 (360)	28 (460)	38 (560)	48 (660)	58 (760)	68 (860)	78 (960)	88 (1060)
Circuit No. 4 90 ms	4 (190)	11 (290)	21 (390)	31 (490)	41 (590)	51 (690)	61 (790)	71 (890)	81 (990)	91 (1090)
Circuit No. 5 120 ms	6 (220)	14 (320)	24 (420)	34 (520)	44 (620)	54 (720)	64 (820)	74 (920)	84 (1020)	94 (1120)
Circuit No. 6 150 ms	8 (250)	17 (350)	27 (450)	37 (550)	47 (650)	57 (750)	67 (850)	77 (950)	87 (1050)	95 (1150)
Circuit No. 7 180 ms	10 (280)	20 (380)	30 (480)	40 (580)	50 (680)	60 (780)	70 (880)	80 (980)	90 (1080)	97 (1180)
Circuit No. 8 210 ms	13 (310)	23 (410)	33 (510)	43 (610)	53 (710)	63 (810)	73 (910)	83 (1010)	92 (1110)	98 (1210)
Circuit No. 9 240 ms	16 (340)	26 (440)	36 (540)	46 (640)	56 (740)	66 (840)	76 (940)	86 (1040)	94 (1140)	99 (1240)
Circuit No. 10 270 ms	19 (370)	29 (470)	39 (570)	49 (670)	59 (770)	69 (870)	79 (970)	89 (1070)	96 (1170)	100 (1270)

1 — Order holes shoot

100 — Shooting time, ms

Du Pont MS delay caps used		Du Pont MS delay caps used	
Cap No.	ms	Cap No.	ms
4	100	15	600
8	200	16	700
10	300	17	800
12	400	18	900
14	500	19	1000

Fig. 5.20 Ten rows, 100 holes, 100 delays; minimum time between delays, 10 ms. (Set blasting machine on 30 ms. Four circuits energized before first cap shoots.) (*Research Energy of Ohio, Inc.*)

SIX

THEORY OF BREAKAGE

PURPOSE OF BLASTING

The purpose of blasting is to convert rock from one solid piece of geologic material to several smaller pieces capable of being moved or excavated by available equipment. To accomplish this there are two major factors to consider: (1) fragmentation and (2) movement, or throw. Both of these must match the requirements of the individual rock project. Underground blasting, for example, requires greater fragmentation than surface blasting because of the size of equipment that can be used and the difficulty of access. If the fragmentation is too large, the equipment will be unable to excavate the rock; and if the fragmentation is too small, the blasting has accomplished more than necessary and therefore the cost is higher than it should be. If the blast creates insufficient movement, the blasted rock will be in such a position that it is broken but not moved enough to excavate easily. If, however, there is too much movement of the rock, there may be damage to surrounding property or even injury to personnel.

THEORY OF BREAKAGE

To achieve the desired fragmentation and movement the engineer must be acquainted with the theory of breakage in order to design the blast that will produce the desired results at a minimum cost. The breaking of rock involves two basic processes, radial cracking and flexural rupture.

Rock is stronger in compression than in tension. Some rocks may have compressive strength approaching 25,000 lb/in^2 and yet be unable to withstand tension above 2500 lb/in^2. Considering this fact it would seem reasonable that the easiest way to break rock is to subject it to a tensile stress greater than its ultimate strength in tension. However, this process can be complicated by the realization that rock formations are rarely homogeneous: the rock formation in one blast area may contain different types of rock, and even more important is the change in density caused by dirt seams and faults.

These changes in density can make it difficult to determine the proper location for the explosives to subject the rock to tensile stresses.

When the blast first occurs the explosion creates a sudden application and quick release of high pressure, sending a shock wave through the rock from the borehole (after first crushing a small amount of rock immediately around the borehole). This compressive shock wave travels from the borehole throughout the entire rock mass as an elastic wave with its speed a function of the rock density—the denser the rock, the faster the wave travels. The wave travels the distance from the borehole to the nearest free face (this distance is the "burden") and then is reflected back to the borehole. If the shock wave reaches a point where there is a change in density, a portion of the wave will return to the borehole and the remainder will continue through the different material in a weakened state. (See Figure 6.1.)

At every change in density some of the impulse energy is reflected and refracted back to the source while the balance of the wave continues through the different material. Particles will begin to spall away from the face if the wave emits energy in excess of the cohesive strength of the rock. This spalling, or breaking away of the rock particles, begins on the free face or at the change in density of material and moves back toward the source, as does the reflected wave.

Proper fragmentation results when there is enough force in the compressive wave to travel to the face and back, overcoming the tensile strengths of the medium through which it travels.

It has been estimated that only 3% of the explosive energy is used in the compression wave in the burden area. Along the face the outermost edge is stretched in tension, which causes cracks.

If the energy released is not high enough to travel to the face and return to the source, boulders will be found in the muck pile around the borehole. However, if there is excessive energy, it will cause additional throw and overbreak.

Boulders on the surface of the muck pile can be caused when the stress wave is reflected because of density changes or seams, or overbreak from a

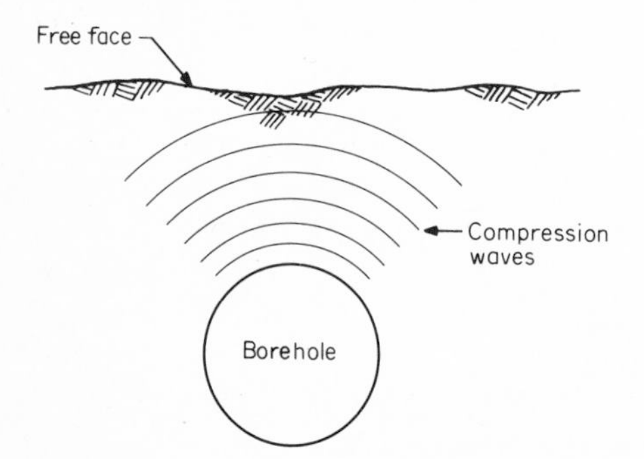

Fig. 6.1 Compression waves.

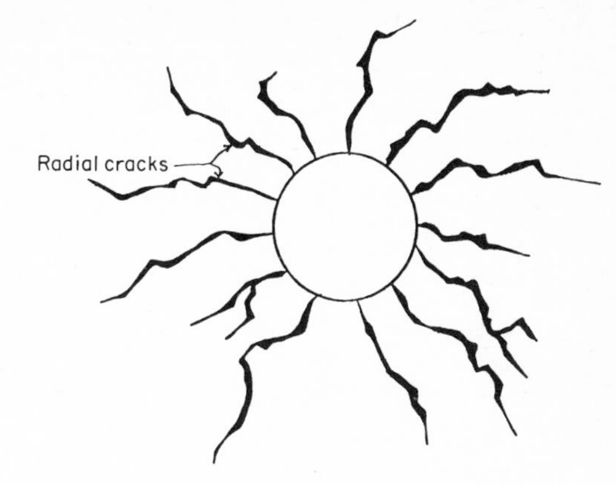

Fig. 6.2 Radial cracks.

previous blast. Instead of being fragmented by the blast, these portions are simply pushed or rolled out of the face.

Radial Cracks

The compression wave is short-lived and will normally enlarge only short radial cracks that radiate out from the borehole axis (see Figure 6.2). If the rock is homogeneous and therefore lacking many radial cracks, the compression wave will help form these radial cracks. In most rocks, however, the cracks already exist. In actuality, the shock wave primarily helps magnify the existing cracks and directs the energy that follows.

Flexural Rupture

The second process in breaking rock is flexural rupture,[1] i.e., bending the rock to the point where the outside edge, the side in tension, breaks. Flexure, or bending, is caused by the rapid expansion of the gases in the borehole. (See Figure 6.3.) This gas expansion exerts pressure against the cylinder walls (the borehole). The sustained gas pressure drives the radial cracks through the burden to the free face and then causes the rock to displace in the direction of least resistance. This gaseous pressure applies the force that is necessary to cause a flexural rupture of the rock and is responsible for the fracture of the rock in a direction perpendicular to the borehole axis (see Figure 6.4). The rock will break following the naturally occurring planes of weakness: joints, cracks, and seams.

Flexural rupture of the rock is analogous to the bending and breaking of a beam. Gas is the major component necessary for the flexure and therefore the primary component of the fragmentation. By treating the section of rock as a concrete beam, with depth equal to the burden and length equal to the bench

[1] C. J. Konya and R. L. Ash, Flexural Rupture: A New Theory on Rock Breakage by Blasting, *Proc. Intern. Conf. Explosives and Blasting Techniques*, WIFI, Linz, Austria, 1975.

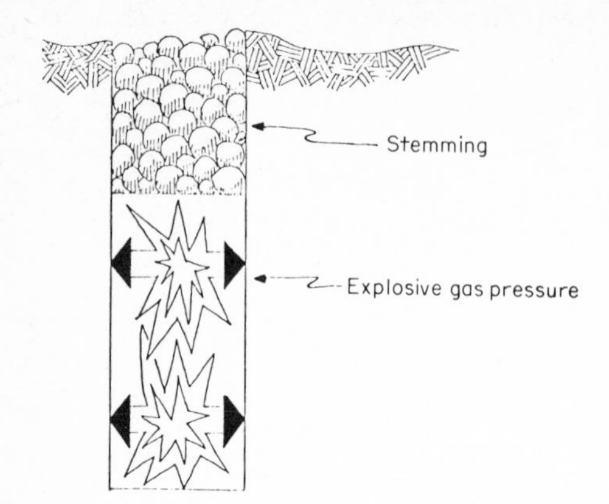

Fig. 6.3 Explosive gas pressure exerting force against borehole walls.

height, it is obvious that to break the beam will require movement, or displacement, in addition to cracking. When the explosive has been detonated and the radial cracks have been expanded, the gas starts the movement by putting a compressive pressure against the wall of the borehole. This pressure, as with a beam, will have the least resistance to flexure at the center of the span. The action begins at the location of the primer; therefore, the largest movement will favor the primer side of the center of the explosive column. As the expansive force of the gas moves against the beam, the beam bends. The bending of the beam creates tensile stresses at the free face of the rock, which break the rock at the locations of the weakness planes caused by either changes in material or the radial cracks caused by the initial compression wave.

The principle of flexural rupture explains the relationship of the length of the borehole to the burden. Burden is the distance from the borehole to the nearest free face. Through the understanding of the effects of flexural systems and varying movements it is clear that the deeper the borehole, the greater the permissible burden and borehole spacing to achieve desired results.

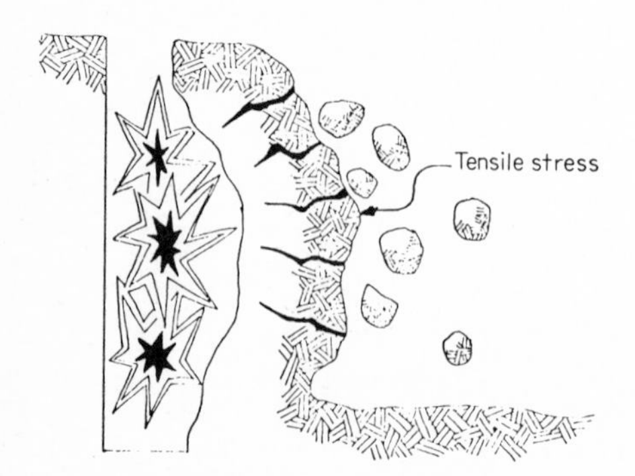

Fig. 6.4 Bending causes tensile stresses in rock.

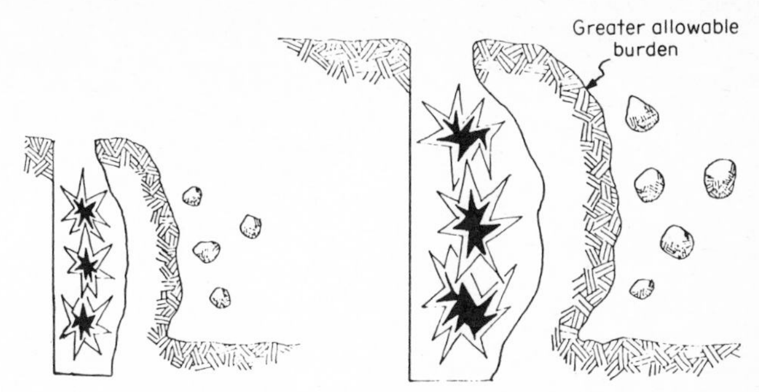

Fig. 6.5 The greater the depth of the borehole, the larger the allowable burden.

Length

The length of the borehole has a direct effect on the amount of burden—the thickness of the beam—that can be broken by flexure: the longer the beam—the depth of cut—the greater the burden that can be broken (see Figure 6.5). Therefore, the shorter the cut, the smaller the burden that can be moved. This principle can be demonstrated with the engineering formula for flexural deflection in beams. In computing the bending moment for a beam, assuming a uniform load, we have the formula:

$$M = \frac{wl^2}{8}$$

where M = the bending moment
w = the load
l = the length

You will note from the above formula that the length is squared. Therefore, if the length of the beam is doubled it increases the bending moment by a factor of 4.

Research[2] indicates that the best ratio for the length of the borehole to the burden is 3 : 1. That is, for maximum fragmentation and minimum overbreak the height of face should be 3 times the burden distance.

Stemming

Stemming is nonexplosive material that is placed in the borehole between the top of the explosive column and the collar of the hole. Stemming can consist of sand, drill fines, or gravel.

It is important that all holes be stemmed to help confine or delay the escape of expansive gases through the top of the borehole, so that the explosive's efficiency is increased. Increasing the explosive's efficiency reduces the

[2] R. C. Ash and N. S. Smith, Changing Borehole Length to Improve Breakage: A Case History, *Proc. Second Conf. Explosives and Blasting Techniques*, Society of Explosives Engineers, Louisville, 1976.

amount of explosive required. If the gases are not properly confined the results can be flyrock, increased ground vibration, and air blast. The reduction in efficiency causes poor fragmentation and boulders.

When a borehole is improperly stemmed, or not stemmed at all, the resulting action is called "rifling." Rifling gets its name from the action of blowing the stemming much like a rifle. The action desired in blasting is the opposite of that with a rifle. A rifle is designed to eject a projectile when the gunpowder explodes. However, if the round had a very powerful explosive, say nitroglycerin dynamite, instead of gunpowder, the projectile would be ejected out of the barrel but the rifle would blow apart. If the barrel were plugged, the projectile would not be ejected; instead, the only action would be the destruction of the rifle. This is the effect desired in a borehole: the destruction of the borehole, not the launching of projectiles. If the stemming is insufficient a projectile will be launched, in the form of either some or all of the stemming or, when no stemming is used, pieces of rock from around the collar of the hole.

The best type of stemming to use is angular-type stemming, such as peastone. Not only is peastone convenient to use because of its availability and its density causes it to settle faster than sand in wet holes, but because of its angular shape it is also the best type of stemming to resist ejection from the borehole. Angular stemming tends to bridge across the borehole as the gas expansion pushes against it. This pushing by the expansive gases causes the stemming to move slightly, wedging itself against the wall of the borehole, with the high pressure causing the angular points to gouge deep into the wall of the borehole, restricting further movement. Because the high gas pressure is short-lived and pushes against the stemming, there is little gas leakage, even though the stemming material is quite porous.

Tests[3] have suggested that the size of the stemming particle has an effect on the performance of the stemming. A borehole-to-particle size ratio is 17 : 1 appears to be the best. That is, ¼-in (6.3-mm) peastone would be the optimum size for a 4- to 4½-in (101- to 114-mm) borehole.

Detonating Velocity

Earlier theories contended that the detonating velocity of the explosive was the chief factor for consideration. However, researchers have learned that the only value that the detonating velocity plays in blasting performance is the formation of short radial cracks. That is, the faster the detonating velocity, the better its radial cracking ability. However, the faster explosives tend to have less gas production. Therefore, one must be careful not to sacrifice the needed gas production for radial cracking.

[3] C. J. Konya, The Effects of Stemming Consist on Retention in Blastholes, *Fourth Conf. Explosives and Blasting Techniques*, Society of Explosives Engineers, New Orleans, 1978.

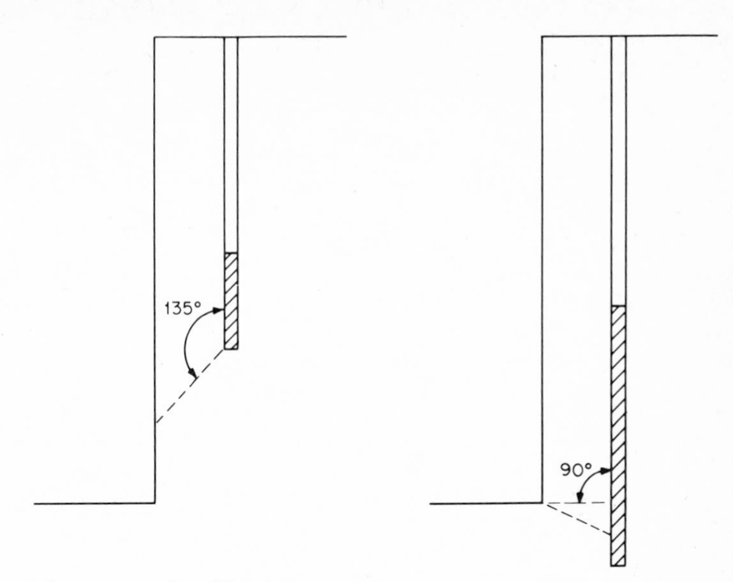

Fig. 6.6 Angles of breakage.

Angle of Breakage

The angle of breakage is the measured angle at which a homogeneous material can be expected to break from the explosive charge to the free face. If the bottom is a free bottom the anticipated vertical angle of breakage is about 135°. However, if it is a fixed bottom, as with bench blasting, the angle should be 90°. (See Figure 6.6.) The horizontal angle of breakage is generally 90°, depending on the burden distance.

SEVEN

BLAST DESIGN

For years, blasting was done on a hit-or-miss basis. The blaster would, on the basis of experience, choose the blast design that would seem likely to give the desired results. Often, in gaining experience, the blaster received undesirable results. Too many blasts proved to be both uneconomical and unsafe. A blaster who always worked in the same geological conditions would eventually arrive at a combination of variables that produced the desired results, although in many cases that method was still not the most economical.

FACTORS AFFECTING DESIGN

There are many factors, or variables, that affect the blast results. The geological factors are out of the blaster's control; however, the blaster may, with a knowledge of the geological factors, place values on the controllable variables to safely achieve the desired fragmentation at minimum cost. These factors are: diameter of the borehole, burden, spacing, stemming, and the design of the delay firing system. These variables are mutually dependent, or related to one another.

Hole Diameter

Generally there are three criteria in determining the borehole diameter to be used: the availability of equipment, the depth of the cut, and the distance to the nearest structure. To reduce the amount of drilling, the blaster will usually use the largest hole that the depth of the cut and the proximity of structures will permit. The maximum borehole diameter that can be effectively used depends on the depth of the hole. Conversely, the minimum depth to which a hole can be drilled is dependent on the diameter and can generally be represented by the formula:

$$L = 2D_h$$

where L is the minimum length of the borehole in feet and D_h is the diameter of the borehole in inches. (For metric, multiply the hole diameter by 25.4 to obtain the minimum hole depth in millimeters.)

The distance to the nearest structure can have a limiting effect upon the diameter of the borehole because of the vibration effect of the blast. The effects of vibration caused by a blast are determined by the amount of explosives detonated simultaneously in relation to the distance to the nearest structure: if the blast is close to structures, the amount of explosives detonated in an instant has to be decreased. There are four ways of decreasing the amount of explosives: to use delay firing devices, to shorten the depth of the cut, to decrease the diameter of the borehole, and to use decking techniques. The four methods may be used individually or in any combination. If the depth of cut is large enough to permit a borehole of any diameter and the distance to the nearest structure is far enough not to pose any concern, then the final determining factor for borehole diameter is the availability and practicality of equipment. To illustrate: If the job consists of a 30-ft (9-m) cut, the depth of the cut permits a large-diameter borehole, but the depth may be too shallow to be economically drilled with a large drill or the job may be too small to warrant the move-in costs. Therefore, the blaster may decide to use a crawler drill for a 3- or 3½-in (76- or 90-mm) borehole, which will cause a decrease in the size of the allowable burden. The lower setup costs may offset the greater cost of drilling more holes.

Burden and Spacing Determination

The burden is the distance from the blast hole to the nearest perpendicular free face. The true burden can vary depending upon the delay system used for the blast; therefore the delay design should be determined before the drill pattern is laid out. In a delay pattern laid out with the holes forming a rectangle, if no adjustment is made in the true and the drill burden, there may be unnecessary drilling.

In Figure 7.1 you will note that the delay number (these are MS delays) is inside the circles representing the boreholes. The first figure shows the drilling burden as the distance from the borehole to the natural free face of the rock. Figure 7.2 shows a delay plan that makes the true burden equal the drilling burden by having all the holes on the same row fire simultaneously.

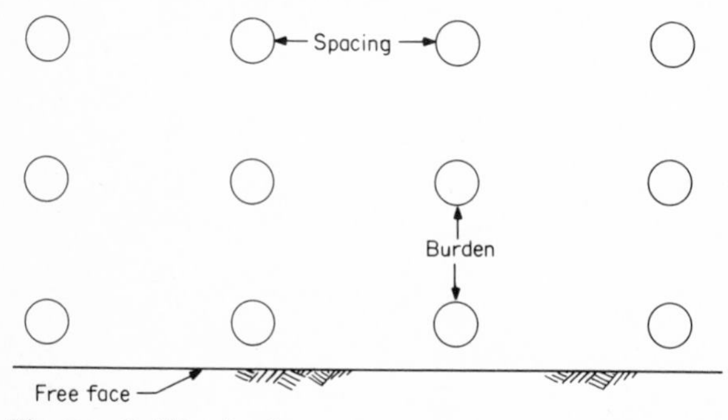

Fig. 7.1 Drilling burden.

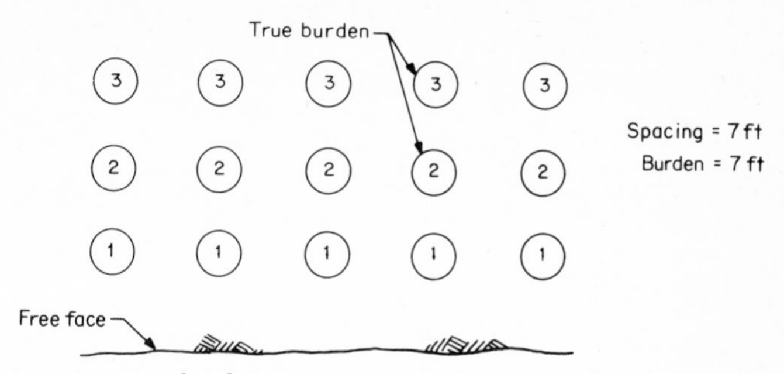

Fig. 7.2 True burden.

However, if the delay pattern is altered, it is obvious that with the use of a delay system the burden and spacing can be moved farther apart, as in Figure 7.3. Burden determination is related to the diameter of the explosives, the depth of the hole (or length of column charge), and the rock and explosive properties. However, early research created simple methods of determining burden.

Andersen Formula

One of the first of the modern blasting formulas was developed by O. Andersen. He developed his formula on the premise that burden is a function of the hole diameter and the length of the hole. The Andersen formula is:

$$B = \sqrt{dL}$$

where B = burden, ft
d = diameter of borehole, in
L = length of borehole, ft

We have learned since this formula was developed that there are more factors involved in burden determination than just hole diameter and length. However, Andersen was quite correct in his assumption that length is a factor in determining the burden. As was discussed in Chapter 6, the length of the burden relative to the depth of the cut has a significant effect on fragmentation.

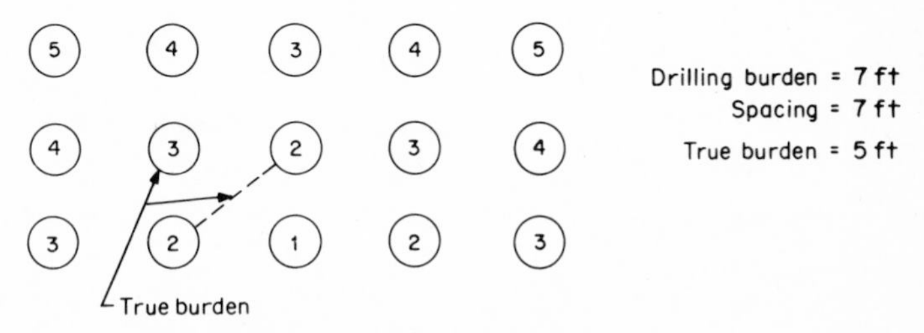

Fig. 7.3 True burden with delay pattern.

Langefors' Formula

Langefors suggested that the burden determination was based on more factors, including diameter of hole, weight strength of explosives, degree of packing, a rock constant, and the degree of fracture.

Langefors' formula for burden determination is:

$$V = (d_b/33)\sqrt{\frac{Ps}{\bar{c}f\,(E/V)}}$$

where V = burden, m
d_b = diameter of drill bit, mm
P = the degree of packing = 1.0 to 1.6 kg/dm^3
s = weight strength of explosive (1.3 for gelatin)
$\bar{c}$ = rock constant, generally 0.45
f = 1 degree of fraction, for straight hole = 1
E = spacing
E/V = ratio of spacing to burden

Using the following metric values the formula will yield the burden necessary.

$$P = 1.25 \text{ kg/dm}^3$$
$$s = 1$$
$$\bar{c} = 0.45$$
$$f = 1$$
$$E = 1.25\ V$$
$$V = (d_b/33)\sqrt{\frac{1.25 \times 1}{0.45 \times 1 \times 1.25}}$$
$$V = d_b/22$$

Therefore, if

$$d_b = 75.9 \text{ mm}$$
$$V = \frac{75.9}{22}$$
$$V = 3.5 \text{ m} = 11.3 \text{ ft}$$

With this formula Langefors is showing a relationship that determines the ratio of bit size to the burden based on the effects of the various elements, or givens, in the formula.

Konya Formula

Currently the best formula for burden determination is one developed by C. J. Konya. This formula uses the diameter of the explosives in relation to the specific gravity of the explosive and of the rock.

The Konya formula is:

$$B = 3.15\,D_e\sqrt[3]{\frac{SG_e}{SG_r}}$$

where B = burden, ft
D_e = diameter of the explosive, in
SG_e = specific gravity of the explosive
SG_r = specific gravity of the rock

Applying the formula to a 3-in hole using AN/FO, we can calculate the following burden:

$$D_e = 3 \text{ in}$$
$$SG_e = 0.85$$
$$SG_r = 2.7$$
$$B = 3.15(3) \sqrt[3]{\frac{0.85}{2.7}}$$
$$B = 6.4 \text{ ft}$$

For metric:

$$B = 160 \times 10^{-6} D_e \sqrt[3]{\frac{SG_e}{SG_r} V_e^2}$$

Length-to-Burden Ratio You will note that the Konya formula does not take into consideration the depth of cut. Therefore, with a combination of the Konya formula and trial and error, we will calculate the most efficient burden, relative to the variables in the formula, and maintain the optimum length-to-burden ratio (L/B) of 3.

Using the same givens and adding a length of borehole of 25 ft (7.62 m), we apply the L/B test:

$$\frac{L}{B} = \frac{25}{6.4} = 3.9$$

This L/B value is too great. Therefore, we must adjust one of the variables. The variables that can be changed are the specific gravity of the explosive (by changing the type of explosive) and the diameter of the explosive (by changing the diameter of the borehole). The latter is the most desirable: because the ratio is too high we can see that the burden can be increased, which will decrease the number of holes. Also, to find an explosive with a lower specific gravity than 8.5 (that of AN/FO) would be difficult. Therefore, the cost savings that result from decreasing the number of holes and staying with the relatively inexpensive AN/FO make increasing the burden the best alternative.

With a new set of variables let's again calculate the burden by the Konya formula. Increasing the diameter of the borehole and the diameter of the explosive, we have:

$$D_e = 5$$
$$SG_e = 0.85$$
$$SG_r = 2.7$$
$$B = 3.15(5) \sqrt[3]{\frac{0.85}{2.7}}$$
$$= 10.7 \text{ ft}$$

This gives us a burden distance of 10.7 ft. Applying the L/B test we get:

$$\frac{L}{B} = \frac{25}{10.7} = 2.3$$

This ratio is too small. We must again change the diameter of explosive to obtain the optimum L/B of 3.

The new variables are:

$$D_e = 4$$
$$SG_e = 0.85$$
$$SG_r = 2.7$$

Therefore:

$$B = 3.15(4)\sqrt[3]{\frac{0.85}{2.7}} = 8.5 \text{ ft}$$

The burden is equal to 8.5 ft. Applying the L/B test we find a ratio of:

$$\frac{L}{B} = \frac{25}{8.5} = 2.92$$

This ratio of 2.92 is good. One cannot expect to obtain a ratio of exactly 3.

Therefore, the burden for our blast with a length of face of 25 ft (7.6 m) is 8.5 ft (2.6 m). This is with the drilling of a 4-in (102-mm) hole and with ammonium nitrate.

It must be noted that AN/FO, being a prill (free-flowing) type of explosive, fills the entire column diameter. However, if one used a stiff cartridge explosive the hole diameter would have to exceed the diameter of explosive. The L/B ratio of 3 is the most desirable; however, there will be many occasions where other conditions will make the use of this ratio difficult and impractical.

Spacing Determination

The spacing is calculated in relation to the burden length: i.e., it is necessary to complete the burden calculations before determining the spacing. Spacing

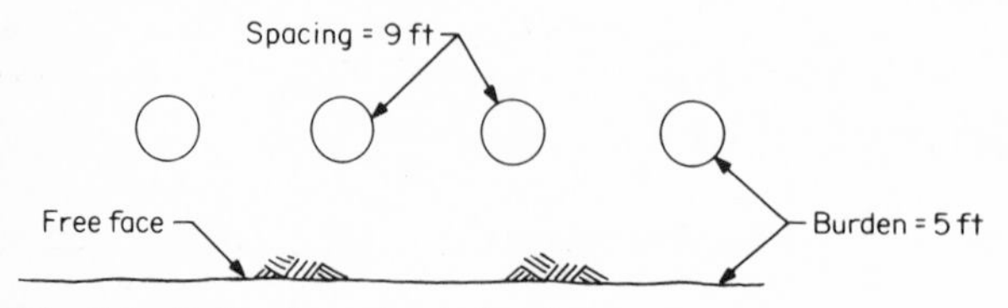

Fig. 7.4 Ratio of spacing to burden.

is the distance between blast holes fired, on the same delay or greater delay, in the same row. (See Figure 7.4.)

For a single-row instantaneous blast, the spacing is usually 1.8 times the burden; i.e., for a burden of 5 ft (1.5 m) the spacing would be 9 ft (2.7 m). For

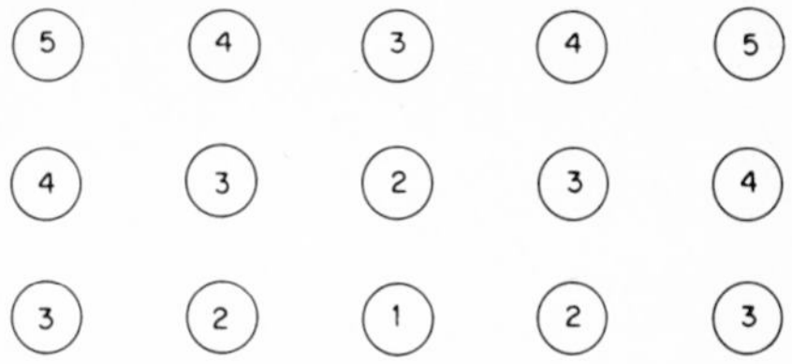

Burden (true) = 5 ft
Burden drillers = 7 ft
Spacing drillers = 7 ft

Fig. 7.5 Delay blasting with square pattern.

multiple simultaneous (same-delay) blasting where the ratio of length of borehole to burden (L/B) is less than 4, the spacing can be determined by the formula

$$S = \sqrt{BL}$$

where B = burden, ft
L = length of borehole, ft

That is, as in the previous example, with a length-to-burden ratio of 3 the spacing would be:

$$B = 8.5 \text{ ft}$$
$$L = 25 \text{ ft}$$
$$S = (8.5 \text{ ft} \times 25 \text{ ft})^{1/2} \text{ or } \sqrt{(8.5)25 \text{ ft}^2}$$
$$S = 14.6 \text{ ft}$$

If the length-to-burden ratio is greater than 4, then the spacing is twice the burden. Therefore, if the L/B is 5 the spacing is determined by $S = 2B$. For example, if the burden were equal to 7 ft (2.1 m), the spacing would be:

$$S = 2(7) \text{ ft}$$
$$= 14 \text{ ft}$$

again showing the relationship of the length of borehole to the burden and spacing dimensions.

A bench height of 10 ft (3 m) with a burden of 5 ft (1.5 m) would require a spacing $S = (5 \times 10)^{1/2} = 7$ ft (2.1 m); whereas a bench height of 30 ft (9.14 m) would require a spacing $S = 2B = 2 \times 5 = 10$ ft (3 m).

In blasting with delay caps the spacing is approximately $1.4B$ for a square pattern with the initiation sequence shown in Figure 7.5. Therefore, in determining the spacing for delay blasting, the delay design must be developed so that the drill pattern can be such as to make optimum use of the spacing length.

Stemming

As discussed earlier the length of the hole has a great effect on the total blast design. The length of the hole is a factor in determining the diameter of the borehole, the burden, and the spacing. Generally the completed borehole is not filled with explosives: except for some underwater blasting, the top of the borehole is not loaded with explosives. If the entire hole were loaded the gases would escape from the top of the hole, causing flyrock and preventing the gaseous buildup from obtaining adequate rock breakage. Therefore, there must be some substance put into the top of the borehole to prevent the gases from escaping prematurely. As discussed in Chapter 6 this substance is called "stemming."

Generally the amount of stemming required will range from $0.7B$ to $1B$. Therefore, in the case of burdens of 8 ft (2.4 m) the stemming will be somewhere between 5.6 ft (1.7 m) and 8 ft (2.4 m). That is, the ratio of stemming to burden can vary through this range depending on existing conditions. The condition most likely to affect the amount of stemming is the structural integrity of the material in the area of the borehole collar. If the material is very competent, as is a homogeneous granite, the stemming will approach $0.7B$. However, if the material is fractured rock with many fissures and dirt seams, the required ratio of stemming to burden will be approximately 1. In material other than rock, such as overburden, the ratio of stemming to burden will be even greater. Overburden is generally treated in a 2 : 1 ratio over rock; that is, 2 ft of overburden is approximately equal to 1 ft of rock for stemming purposes. Therefore, if there is 4 ft (1.2 m) of overburden on a shot that requires a burden of 7 ft (2.1 m), the stemming is calculated as follows.

Using the ratio of $0.7B$, the stemming would ordinarily be 5 ft (1.5 m). However, with 4 ft (1.2 m) of overburden, for stemming purposes equal to 2 ft (0.6 m) of rock, the actual amount of stemming is 7 ft (2.1 m). Another way of expressing it is to compute the required stemming and then add one-half the overburden depth. Thus:

$$\text{Stemming} = 0.7B + \frac{\text{OB}}{2}$$

where B = burden
OB = overburden

Therefore

$$0.7(7)\text{ ft} + \frac{4\text{ ft}}{2} = 7\text{ ft}$$

When the terrain or elevation of the formation changes drastically, the amount of stemming will vary with each hole. The stemming ratio is applied

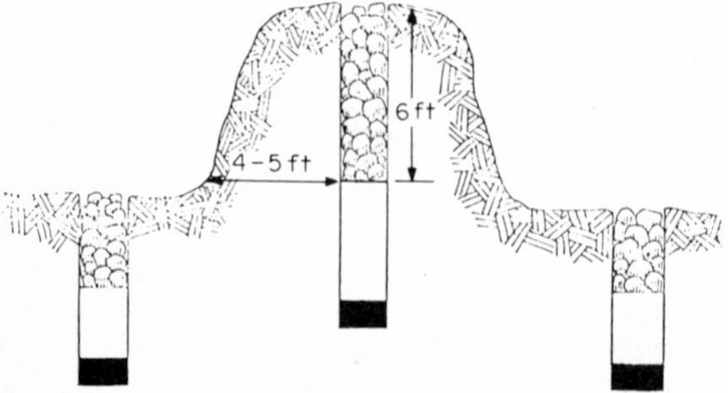

Fig. 7.6 Depth of stemming is figured by determining the burden.

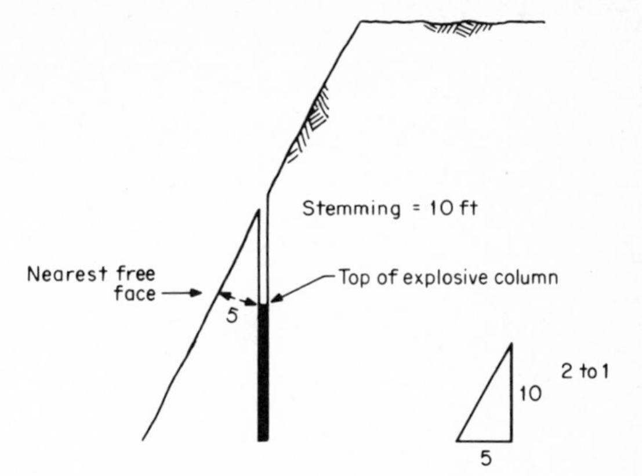

Fig. 7.7 To compute the stemming for a side hill cut, use a right triangle.

to the distance from the top of the explosive column to the nearest free face. In Figure 7.6 the burden is 4.5 ft (1.37 m), and therefore the calculated stemming will be 4.5 ft (1.37 m). However, to use 4.5 ft (1.37 m) of stemming on such a slope would cause undesirable flyrock. Therefore, the borehole will require 6 ft (1.8 m) of stemming to maintain a distance of 4.5 ft (1.37 m) to the nearest free face.

This type of situation is more likely to exist on side hill cuts, i.e., shots that are on a steep side hill. For example, if the stemming for a particular shot is determined to be 5 ft (1.5 m) on a side hill with a 2 : 1 slope, the actual depth of stemming in the borehole is calculated by determining at what depth in the borehole the distance to the free face equals 5 ft (1.52 m). (See Figure 7.7.) As can be seen in Figure 7.7, by deriving a right triangle, the depth in the borehole for stemming can be determined.

Subdrilling

To ensure that the blasting provides adequate fragmentation to the desired grade, it is necessary to drill below the desired grade. This subdrilling is

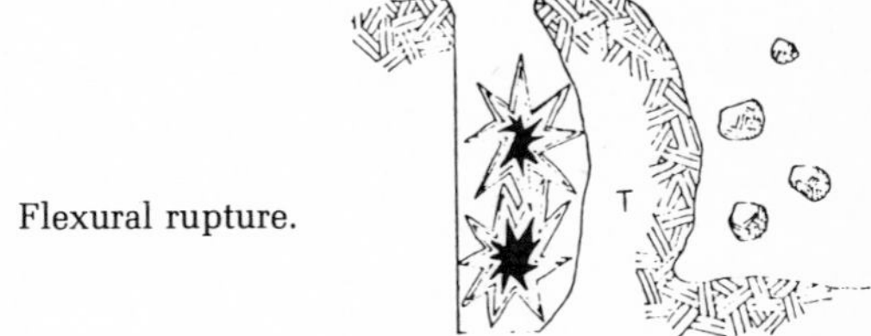

Fig. 7.8 Flexural rupture.

necessary because of the nature of rock breakage: when the explosive is detonated the rock at the bottom of the borehole is the most difficult to break, since it is confined. Figure 7.8 demonstrates flexural rupture. The center of the face (T) is subject to the maximum tensile stresses. It is advisable to

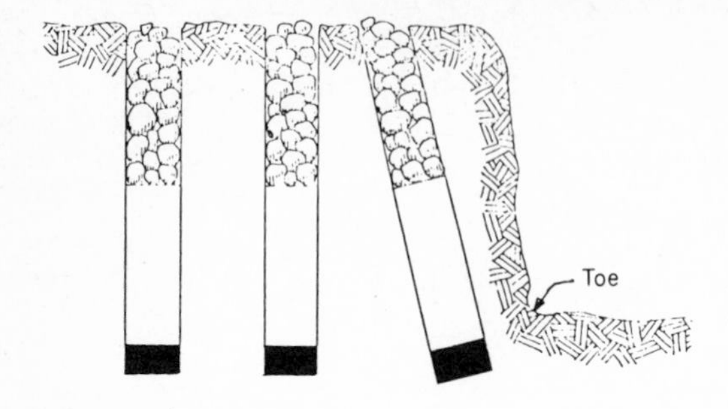

Fig. 7.9 Hole drilled into toe to aid breakage.

subdrill to a depth of at least 0.3 to 0.5 times the burden below the desired elevation to increase the magnitude of the tensile stress at floor level. The subdrilling will vary depending on the type of rock. However, when there is a question, it is better to drill more than necessary rather than have high bottom, which requires expensive secondary work.

The same principle is involved in removing the toe. In Figure 7.9 it is necessary to drill a hole angled below the toe to aid in the bending. If the toe is blasted with an earlier delay than the rest of the shot, the first row of holes can develop flexure. If the angle hole, or looker, is not drilled below the toe, the rock in that area will not exceed the ultimate tensile strength and failure will not occur at grade. This is the result of the geometric properties of the toe: the toe, if not put into a position of tension, will remain in place unless the *compressive* strength of the rock is exceeded (see Figure 7.10). The hole drilled at an angle into the toe will shift the breakage geometry of the toe enough to allow it to bend, converting the stresses from compressive to tensile.

Delay Blasting

As has been discussed previously, there are advantages to blasting with delay firing systems. To reiterate, delay blasting makes possible the reduction of blasting vibrations, helps control the direction of throw, reduces the explosive requirements, and permits the extension or increasing of the burdens and spacings.

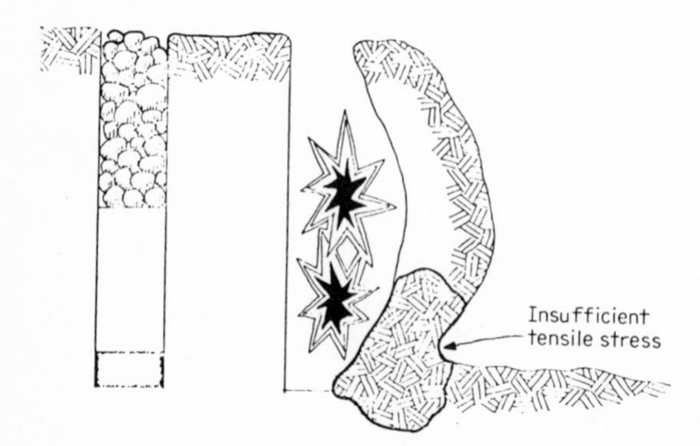

Fig. 7.10 Toe without hole does not develop sufficient tensile stress.

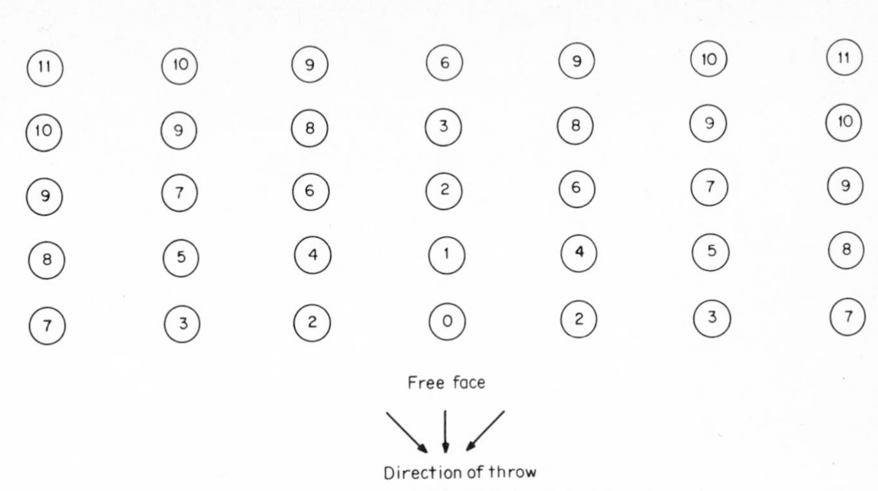

Fig. 7.11 Delay pattern that creates considerable back break.

The delay pattern that is used depends on the desired results and the existing conditions. The earliest firing delay is at the point on the shot that will be the direction of throw; except in cases where presplit is being fired with the shot, the first delay to be fired is in the desired direction of throw either at a natural free face or one being produced by the delay design. If the delays are used primarily to reduce vibration, then the vibration may be reduced, but this reduction may affect back break; i.e., the delay pattern may cause excessive back break because of the many independent detonations along the excavation line (the back row of holes). The requirement is 1 to 2 ms per foot of burden to allow time for the rock to move. (See Figure 7.11.)

Other than working in areas where vibration is a problem, the primary emphasis of delay design for construction blasting is breakage and direction of throw, or muck pile location. Probably the most common type of delay pattern is called the "V pattern." (See Figure 7.12.) This pattern leaves a muck

Fig. 7.12 V delay pattern.

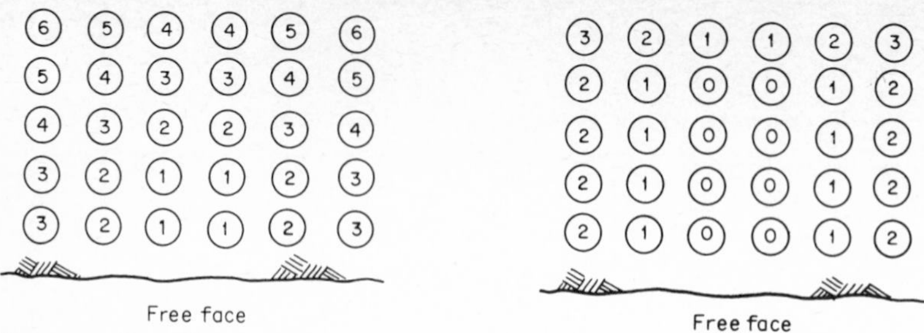

Fig. 7.13 Horseshoe pattern.

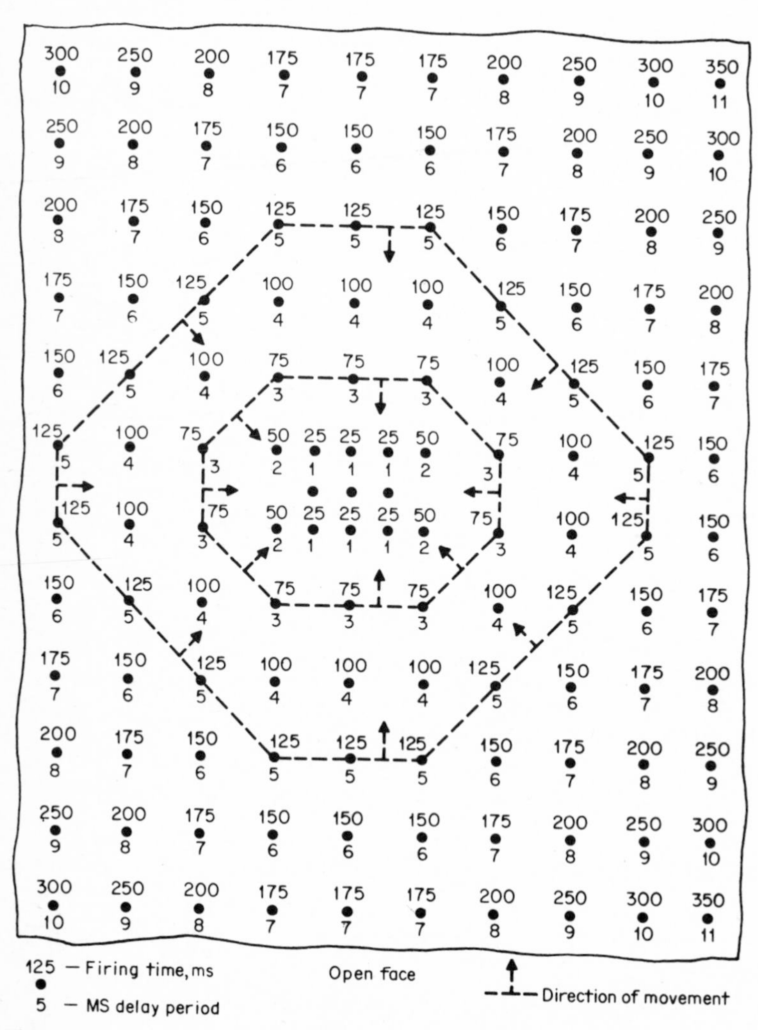

Fig. 7.14 Typical sinking blast pattern using MS delays. The center plug is pulled using nine holes with the burden and spacing about one-half that used in the remaining portion. (*E. I. du Pont de Nemours and Co., Inc.*)

pile extending from the front of the face to the back of the shot, with the muck pile placed high and to the center. Figures 7.11 and 7.12 demonstrate a pattern with the center holes firing first and the remainder firing in a V sequence. These tend to leave a flatter muck pile than the horseshoe or U pattern. With the horseshoe pattern the direction of throw is toward the center and the free face (see Figure 7.13).

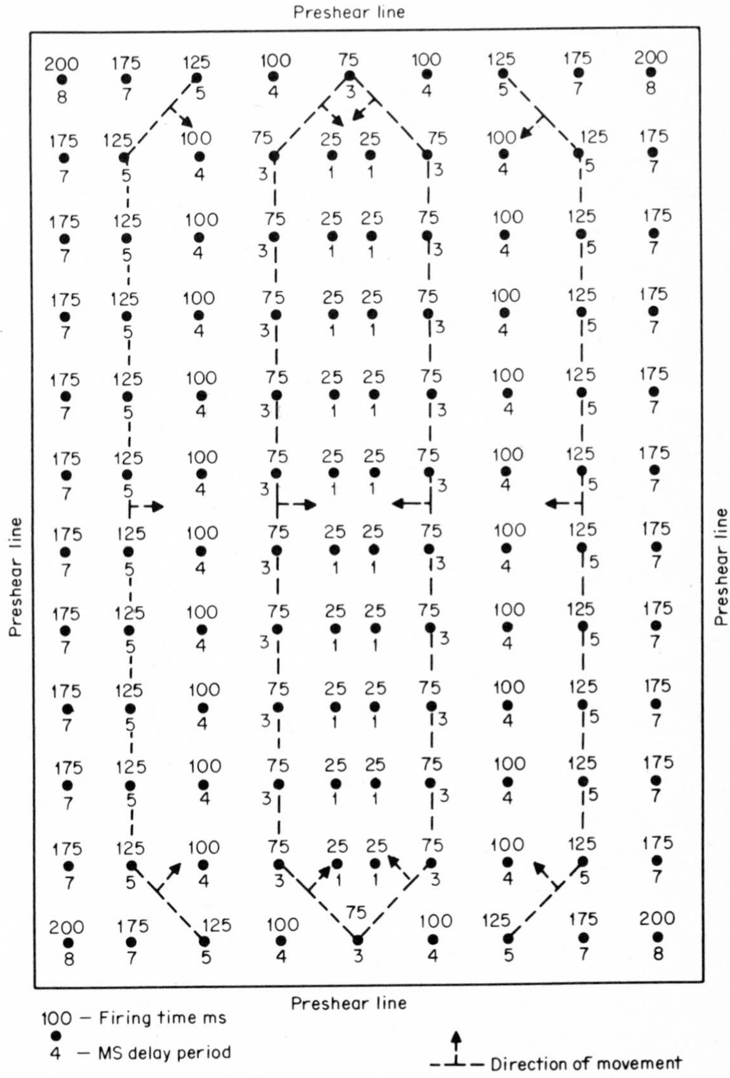

Fig. 7.15 Another sinking blast using MS delays. The two center rows are drilled with close burdens using normal spacing and are fired first. The remaining rows are delayed. (*E. I. du Pont de Nemours and Co., Inc.*)

For sinking cuts, where there are no vertical free faces to shoot to, the shot must be delayed from the center, pulling the four sides of the shot toward the middle. Designs of this type require higher powder factors and cause more vibration. They also, as with shaft-sinking rounds, require short burdens and spacings and greater subdrilling. As with shaft sinking, the center must be pulled first to form a face to which the rest of the shot can break. The burdens and spacings required for the center of the shot are equal to approximately half those required by the remainder of the shot. Figure 7.14 shows a typical sinking-cut pattern where the center plug is pulled first with the cut in a pyramid shape.

Another method of sinking cut is to drill the cut with very short burdens and normal spacing and to have the longitudinal center fire first. This will leave the muck pile in the center of the shot, laying it longitudinally like a windrow. (See Figure 7.15.)

The occasion may arise where on a side hill cut, it is desirable to have the muck pile favor the upper side of the hill instead of following the natural tendency to be directed to the lower side. This may be chosen because the lower side is a bad area from which to excavate (such as soft or swamp-type ground) or the cut is adjacent to an existing highway and one is trying to minimize the costs of clearing the highway. To accomplish this goal, the shot delay pattern must be so designed as to have the direction of movement toward the uphill side of the shot. This is done with an offset V cut. If vibration is a problem this can be used in conjuction with a sequential timer. (See Figure 7.16.)

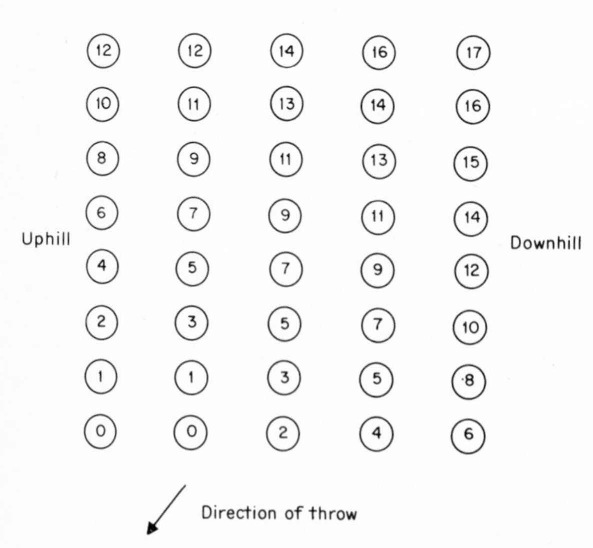

Fig. 7.16 Delay pattern for a side hill cut.

DESIGNING THE BLAST

In this chapter, the various methods of blasting design have been discussed to clarify the interaction of the criteria involved. Let us now design an entire blast. The existing conditions are:

1. Depth of cut = 30 ft (9 m).
2. The dimensions = 50 ft × 80 ft (15 m × 24 m).
3. Vibration is not a problem.
4. Ammonium nitrate is the explosive to be used. Specific gravity is 0.8.
5. Rock type is homogeneous granite. Specific gravity is 2.7.
6. The delay pattern is a V type.
7. Drill availability will match optimum blasting conditions.

We will start the calculations by assuming a 3-in diameter explosive.

$$B = 3.15 D_e \sqrt[3]{\frac{SG_e}{SG_r}}$$

$$B = 3.15(3) \sqrt[3]{\frac{0.8}{2.7}}$$

$$B = 6.3 \text{ ft}$$

The L/B test yields:

$$\frac{30 \text{ ft}}{6.3 \text{ ft}} = 4.76$$

This is too high; by increasing the borehole we can expand this burden and spacing. Trying an explosive diameter of 5 in,

$$B = 3.15(5) \sqrt[3]{\frac{0.8}{2.7}}$$

$$B = 10.5 \text{ ft}$$

$$\frac{L}{B} = \frac{30 \text{ ft}}{10.5 \text{ ft}} = 2.85$$

This is acceptable.

We have established that the burden is 10.5 ft (3 m). From this we now determine the spacing, subdrilling, and stemming.

Spacing:

$$S = 1.4B$$

$$S = 1.4(10.5) \text{ ft}$$

$$S = 14.7 \text{ ft}$$

We'll round this to 15 ft (4.5 m).

Subdrilling:

$$SD = 0.4B \text{ for granite}$$
$$SD = 0.4(10.5) \text{ ft}$$
$$SD = 4.2 \text{ ft}$$

We'll round this to 4.5 ft (1.37 m)—better to be conservative.

Stemming:

$$\text{Stemming} = 0.7B$$
$$\text{Stemming} = 0.7(10.5) \text{ ft}$$
$$\text{Stemming} = 7.4 \text{ ft}$$

In light of these calculations we can now lay out the drill pattern.

As can be seen on the drill pattern (Figure 7.17), we have 35 holes plus overbreak by drilling; i.e., we will have to drill beyond the excavation limits. This occurs more often than not on nonpresplit cuts. However, if the excavation line is critical, then the specifications should relay that information. For the sake of simplicity, assume that there is no penalty for overexcavation.

Now the quantity can be easily calculated:

Total number of holes = 35
Total depth of hole = Height of face plus subdrilling
= 30 ft + 4.5 ft (9.1 m + 1.4 m)
= 34.5 ft (10.5 m)
Total drilling = number of holes times total depth
= 35 × 34.5 ft
= 1208 ft (368 m)

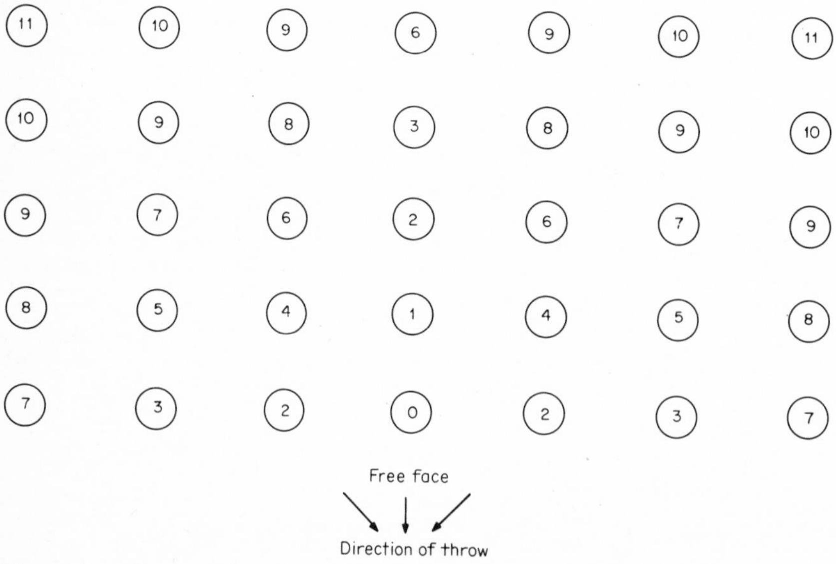

11 10 9 6 9 10 11
10 9 8 3 8 9 10
9 7 6 2 6 7 9
8 5 4 1 4 5 8
7 3 2 0 2 3 7

Free face
Direction of throw

Fig. 7.17 Drill pattern for 35 holes.

To compute the total quantity of rock,

$$6S \times 5B \times \text{total depth} = 90 \text{ ft} \times 52.5 \text{ ft} \times 34.5 \text{ ft}$$
$$= 163{,}013 \text{ ft}^3 \ (4617 \text{ m}^3)$$

In cubic yards, the total quantity of rock is:

$$\frac{163{,}013 \text{ ft}^3}{27 \text{ ft}^3/\text{yd}^3} = 6038 \text{ yd}^3 \ (4617 \text{ m}^3)$$

Total explosives:

$$\text{Hole quantity} = \text{total depth minus stemming}$$
$$= 34.5 - 7.4$$
$$= 27.1 \text{ ft (8.3 m) of explosives}$$
$$\text{Loading density of AN/FO} = 7 \text{ lb/ft (10.4 kg/m)}$$
$$27.1 \text{ ft} \times 7 \text{ lb/ft} = 189.7 \text{ lb per hole}$$

Add one cartridge of 60% semigelatin at 16⅔ lb. Reduce the AN/FO to 180 lb, because the cartridge length takes up 26 in of borehole.

$$180 \text{ lb} + 16\tfrac{2}{3} \text{ lb} = 196\tfrac{2}{3} \text{ lb}$$

Round this to 195 lb per hole.

$$\text{Maximum pounds per delay} = 4 \text{ holes} \times 195 \text{ lb}$$
$$= 780 \text{ lbs (354 kg)}$$

Total explosives required:

$$195 \text{ lb/hole} \times 35 \text{ holes} = 6825 \text{ lb (3096 kg)}$$
$$\text{Powder factor} = \frac{6825 \text{ lb}}{6038 \text{ yd}^3} = 1.13 \text{ lb/yd}^3$$

If this is the first of several shots, we may find that the fragmentation is greater than required. Therefore, because of the delay pattern and the powder factor, we may be able to spread the burdens or spacings or both even farther apart on a subsequent blast. The delay pattern used would tend to place the muck pile too high and toward the center. Also, you may elect to use more delays, thereby reducing vibration. However, to obtain the reduction of vibration, there may be a sacrifice with additional overbreak.

The formulas in this chapter were used to give an idea of how a blast may be planned. It is not recommended that these formulas be applied by unqualified people. These formulas, if used improperly or for the wrong types of conditions, could prove dangerous.

To summarize, the blast design should be done only by qualified, experienced blasting or explosives engineers.

EIGHT

CONTROLLED BLASTING

Controlled blasting techniques are used to control overbreak and to aid the stability of the remaining rock formation. Often there are overbreak penalties that require replacing the overbreak and the loosened adjacent material with concrete. Also, when highways are built through rock cuts, controlled blasting aids in reducing the spreading of rocks onto the highway and contributes to aesthetics. For these reasons many highway, tunnel, and foundation excavation specifications call for some type of controlled blasting. (See Figure 8.1.)

METHODS

There are four methods of controlled blasting, and the one selected depends on the rock characteristics and the feasibility under the existing conditions. The four methods are line drilling, cushion blasting, smooth-wall blasting, and presplitting. The determinants of which of these techniques is used are the rock properties, the ground strength, the diameter of the perimeter holes, the spacings desired, the type of explosives to be used, and the buffer distance available.

All the methods require that holes be drilled on the perimeter of the excavation and that there be a buffer zone between the nearest production blast holes and the perimeter holes. Also, they require precision hole alignment. When the production shot is fired the buffer zone fractures up to the perimeter holes' line but should not pass the line.

Line Drilling

Line drilling provides a plane of weakness to which the rock can break. The line-drilled holes help to reflect the shock wave, reducing the shattering effect of the rock outside the perimeter.

Line-drilling holes, usually due to drilling costs, do not exceed 3 in (76.2 mm) in diameter and are spaced one to four diameters apart, depending on

Fig. 8.1 Typical presplit along a highway.

the rock. Formations with greater occurrences of jointing and planes of weakness require closer spacing between the line-drilled holes. In firm homogeneous formations, the line-drilled holes can be spread out to the four diameters.

Line-drilling holes are not loaded, and the adjacent production holes have less explosive and closer spacings than the remainder of the shot. The width of the buffer zone is 0.5 to 0.75 times the production shot burden distance. The spacing of the adjacent production zone is approximately 0.75 times the normal burden, and only about half the normal explosives are in the borehole. This adjacent row of holes should be loaded with the explosives distributed throughout the hole by a decking technique, with detonating cord. (See Figure 8.2.)

Line drilling requires considerably more drilling than the other controlled blasting methods. Also, the line-drilling method is not very effective in nonhomogeneous formations. In formations that have bedding planes, jointing, and seams, the line-drilling method is not effective in preventing the natural planes of weakness from extending into the finished wall.

Line drilling is more effective when most of the production shot has been blasted, decreasing the mass in front of the face. This will reduce the back-pressure from the shot, thereby cutting reflected pressure waves.

Cushion Blasting

Like line drilling, cushion blasting requires a single row of holes ranging from 2 to 3½ in (51 to 89 mm) in diameter. Unlike line-drilling holes, the cushion holes are loaded with light, well-distributed charges. These holes are fully stemmed between charges, allowing no air gap, and are fired after the production shot has been excavated. The charges should be placed against the production side of the borehole, because the stemming acts as a cushion to protect the finished wall from the shock of the charges when detonated; the larger the borehole, the greater the cushion.

For best results the holes should be detonated simultaneously to achieve a shearing effect in the web; however, if vibration is a problem small delays can be used.

The spacing, generally, is nominally the inches of diameter of the borehole in feet, plus 1. That is, a 2-in hole would require a spacing of 3 ft (2 plus 1).

The holes are string-loaded, with detonating cord used as a downline. The holes are stemmed as they are loaded to maintain the charge distribution; or the charges can be taped to the downline or inserted in spacing tubes ready to be used to maintain the charge distribution, thus allowing the stemming to be done after the hole is loaded. (See Figure 8.3.)

The charge distribution is one cartridge every 2 ft (0.6 m), with the cartridge size a function of the hole size. Therefore, a 3-in (76-mm) hole would have a cartridge of explosive 1½ in (38 mm) in diameter every 2 ft (0.6 m). The bottom of the hole should be loaded approximately 3 times more heavily to move the toe. (See Table 8.1.)

The spacing between the cushion holes should always be less than the burden, preferably 0.8 times the burden distance.

When the perimeter curves, the spacing should be reduced. If possible the loading between each two adjacent cushion holes should be staggered. Although sometimes hard to do in the field, it can be done with relative ease by measuring. (See Figure 8.4.)

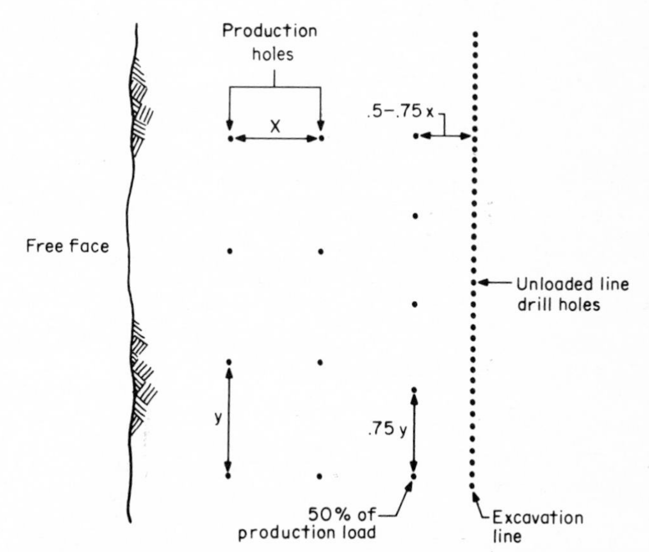

Fig. 8.2 Line drilling: drilling and loading pattern.

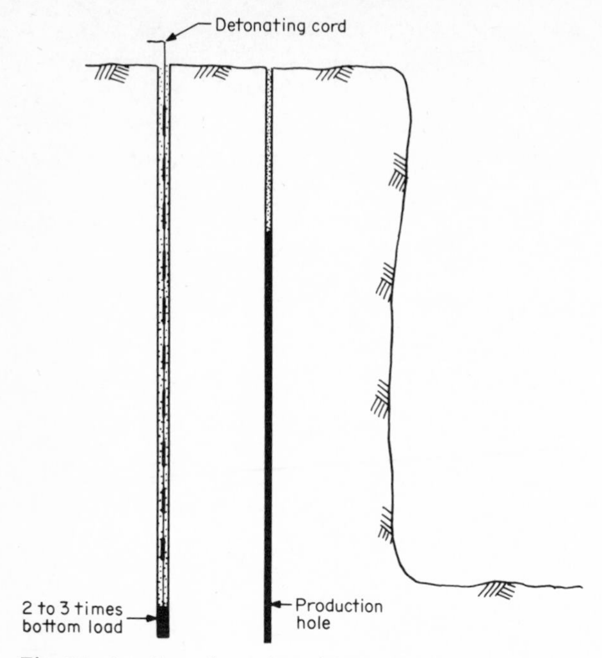

Fig. 8.3 Loading of a cushion-blasting hole.

The easiest way to load the explosive is to use an explosive premade for such use. It comes in ⅞-in (22-mm) and 1-in (25-mm) diameters. (See Figure 8.5.) With tubular coupling, one can have a continuous column of low-density explosives, providing an even distribution. (See Figure 8.6.)

Cushion blasting is not suited for underground applications because of the tough stemming requirements. However, on the surface, it can be used for both incline and vertical holes.

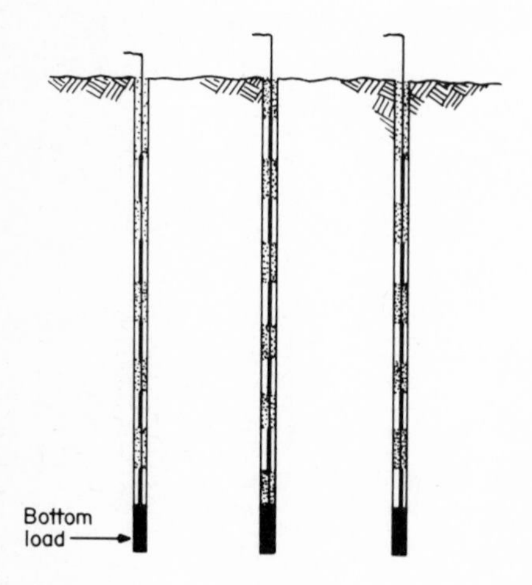

Fig. 8.4 The explosives in adjacent holes should be staggered.

TABLE 8.1 Proposed Loads and Patterns for Cushion Blasting

Hole diameter, in	Spacing,* ft	Burden,* ft	Explosive charge,*† lb/ft
2–2½	3	4	0.08–0.25
3–3½	4	5	0.13–0.50
4–4½	5	6	0.25–0.75
5–5½	6	7	0.75–1.00
6–6½	7	9	1.00–1.50

* Dependent upon formation being shot. Figures given are an average.
† Ideally, dynamite cartridge diameter should be no larger than ½ the diameter of the hole.

SOURCE: "Blaster's Handbook," E. I. du Pont de Nemours and Co., Inc., Wilmington, 1969, p. 403.

Cushion blasting permits a reduction in the number of holes required by line drilling; however, one cannot turn a 90° corner with cushion blasting. It also performs better in nonhomogeneous rock than line drilling. Deeper cuts may be taken, because the larger-diameter holes result in a reduced alignment problem. (See Figure 8.7.)

Cushion blasting requires the removal of the excavated material before firing. This can be costly; there will be production delays, because the excavation of the entire area cannot take place at once. One must partially

Fig. 8.5 Loading with explosives especially designed for cushion and presplit blasting.

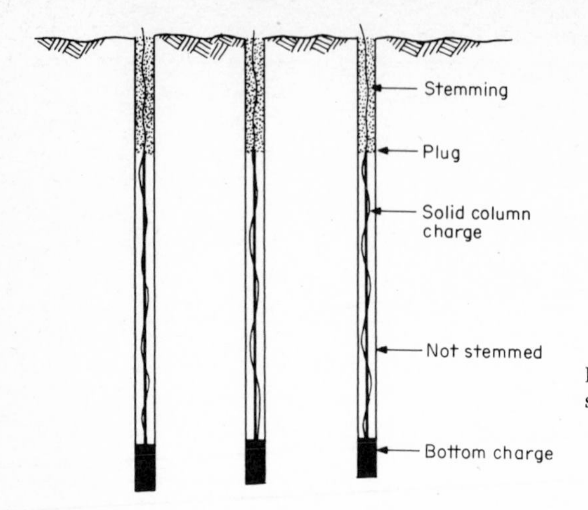

Fig. 8.6 Holes loaded with solid column.

excavate and then return. Another problem with cushion blasting is that sometimes the production shot can break back to the cushion holes, creating redrilling problems and causing loading changes.

Smooth-Wall Blasting

Quite similar to cushion blasting, smooth-wall blasting requires stemming at the collar, but not for the entire length of the hole. Intended for underground

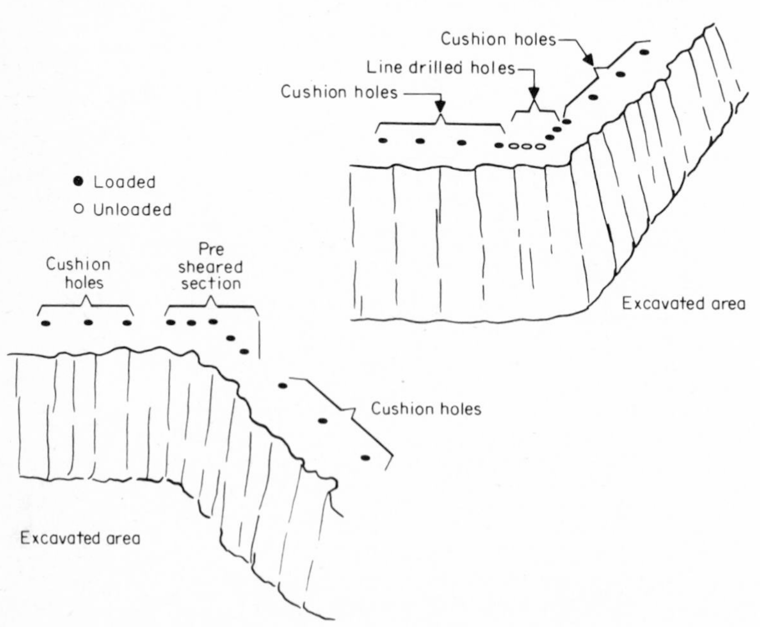

Fig. 8.7 Cushion blasting nonlinear faces. (*E. I. du Pont de Nemours and Co., Inc.*)

use, smooth-wall consists of horizontal holes charged with small-diameter powder cartridges which are fired simultaneously after the lifters. This method works on the same basic principle as cushion blasting; i.e., the excavated material must be removed for the shearing action to work. Therefore, the holes along the lower rib do not obtain as good a result as the back holes because of the muck pile.

The burden is about 1.5 times the spacing, and the collar of the hole must be stemmed to prevent the charges from being pulled from the holes by rock movement caused by earlier delays. (See Figure 8.8.)

This method, like the other methods, works best with homogeneous material, but it will improve the finish of all rocks, thereby reducing the amount of support required. Also, since it reduces the cracking of the remaining rock, the amount of water accumulating in the excavation will not be as great as with techniques other than smooth-wall.

Smooth-wall blasting does require more holes because of the reduced spacings; however, the reduced overbreak can result in savings in excavation and concrete.

Presplitting

Presplitting works by creating a plane of shear in solid rows along the desired excavation before the production blast. The presplit holes may be fired

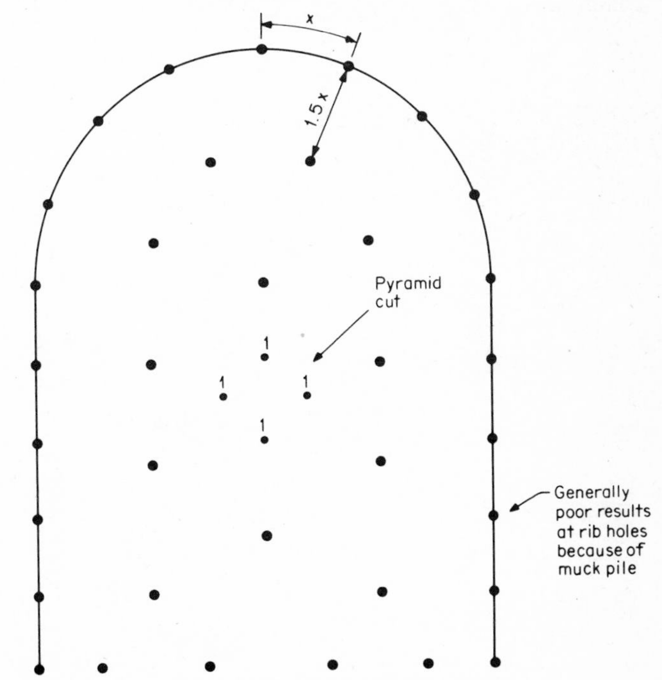

Fig. 8.8 Smooth-wall pattern for a tunnel round.

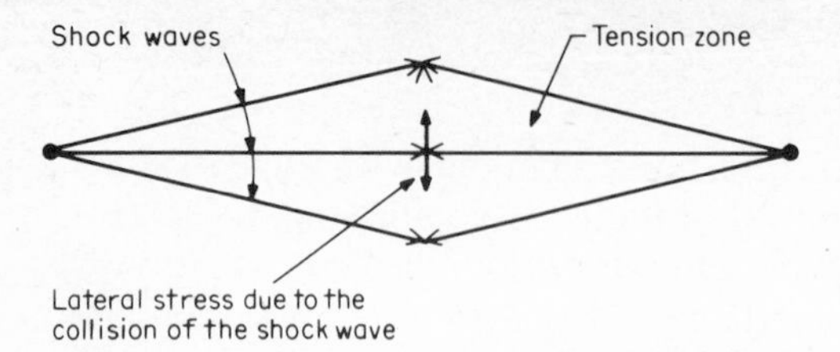

Fig. 8.9 Presplit principle.

Fig. 8.10 Presplitting on incline. (*Hercules Inc.*)

Fig. 8.11 Presplitting in benches. (*Hercules Inc.*)

on an earlier delay than the production shot, or they may be fired before any production shot drilling is done. By providing this sheared plane before the production shot, one not only reduces overbreak but also reduces the vibration.

Presplit holes are 2 to 4 in (51 to 101 mm) in diameter at a spacing equal to one-half the production burden. All the holes are loaded, as with cushion blasting, with a trunk line, but can be delayed.

The presplit theory is that two simultaneously fired holes emit shock waves, which, when they meet within the web, place the web in tension, causing cracks and shearing it. (See Figure 8.9.)

In extremely weathered material, the presplitting may have to be done at very close spacings with a very small amount of explosives.

Presplit holes must be stemmed with an increased bottom charge to move the toe.

The maximum effective depth is approximately 50 ft (15 m), because of hole alignment, and when blasting in tough formations it may be necessary to use uncharged guide holes between the presplit holes. (See Figure 8.10.) Figures 8.10 to 8.15 are photographs of various applications of controlled blasting.

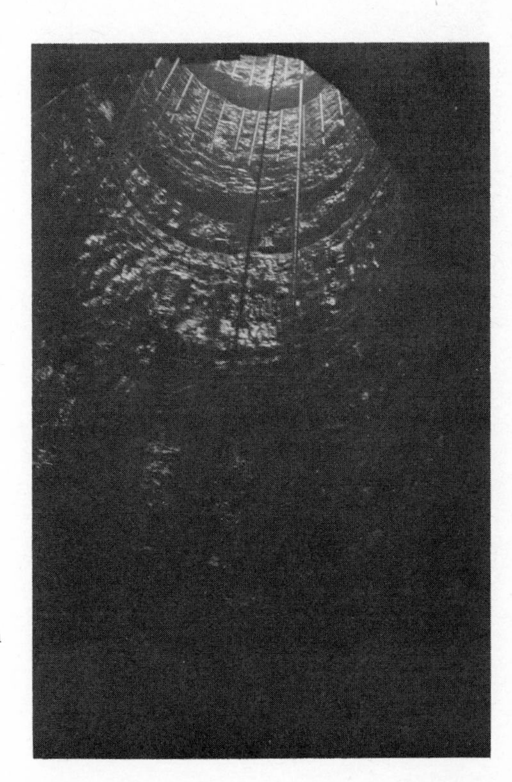

Fig. 8.12 Presplitting for a shaft. (*Hercules Inc.*)

Fig. 8.13 Turning a corner. (*Hercules Inc.*)

Fig. 8.14 Hole paths left by presplitting. (*Hercules Inc.*)

Fig. 8.15 Presplit holes being drilled on an angle. (*Hercules Inc.*)

NINE

BLASTING GEOLOGY

Blasting is affected by the characteristics of the rock being excavated, i.e., the structural integrity, or degree of homogeneity, and the engineering properties. Rocks are often collections of the same materials but in different structural and chemical forms. Therefore, it is important that anyone involved in the fragmentation of rock by explosives have an understanding of the origin of rocks and the structural and engineering properties.

ORIGIN OF ROCKS

There are three primary classifications of rock based on origin: igneous, sedimentary, and metamorphic. Igneous rocks are formed from magma; sedimentary rocks are formed from mineral grains cemented together after being eroded from preexisting rocks; and metamorphic rocks result from changes in igneous or sedimentary rocks caused by heat and pressure.

Igneous Rocks

Igneous rocks are collections of minerals that have been crystallized from molten material. Igneous rocks are formed when magma (molten rock) from within the earth's crust, or mantle, rises to the earth's surface through cracks or fissures in overlying layers of rocks. It is estimated that 95% of the earth's crust is made up of igneous rocks. Magma is any hot molten material that can penetrate rocks of the earth's crust. It is normally a high-temperature mixture, or solution, of silicates containing water vapor and other gases. Magma comes from beneath the earth's crust, where the temperature increases with depth. At a depth of approximately 20 mi (32 km) the temperature is 900–1100°F (482–593°C). This temperature is normally enough to melt rock; however, at that depth the pressure is so great that the melting point of the rock has not yet been reached. For magma to form, there must be an increase in temperature and/or a decrease in pressure.

Magma can come to the surface in either of two kinds of areas. The first type is called low-velocity zones, where the rocks are believed to exist in a melted or partially melted state. Once the rock is melted it comes to the surface by melting or pushing aside overlying rocks. The original composition, or mineral content, is changed because the magma picks up chemicals and minerals from the rock through which it passes. (For a partial list of elements found in rocks, see Table 9.1.)

Magma can also come to the surface where continental and oceanic plates collide. At these points the oceanic plate slides under the continental plate with a friction that creates enough heat for partial melting. The magma rises on the landward side, the upper plate, through tensional cracks in the overlying rock. The melting of the rock from plate movement also causes a change in the chemical makeup of the rock.

Lavas are magma that reach the surface of the earth and flow out over it. This magma cools to form extrusive rocks, which are either of a fine crystalline grain or of a glossy texture. (See Table 9.2 for the classification of igneous rocks.)

Intrusive rocks have a coarse crystalline texture. The most common intrusive rock is granite. Large bodies of intrusive rock deep in the earth's crust are known as "plutonic rocks." Hypabyssal rocks are intrusive rocks that are fairly near the earth's surface that fill in cracks and fissures, forming sheets between the existing layers. These hypabyssal rocks cool more rapidly than the pluton rocks and therefore have a smaller grain. Examples of hypabyssal rocks are felsite and dolerite.

Batholiths are the largest form of plutonic intrusions. They consist of a mass of igneous rock that has been pushed up through the overlying rock strata, causing the overlying rock to be displaced and broken. Because of the disruption in the overlying rock, batholiths are called "discordant intrusions." After the other, displaced rock has eroded away, the batholith will appear on the surface.

If an intrusion does not break the rock layers it is considered a "concordant intrusion." An example of a concordant intrusion is a laccolith, which is an intrusion of more viscous magma.

Extrusive rocks are formed by rapid cooling and therefore do not develop large crystals: they are fine-grained or glossy. The most common source of

TABLE 9.1 Elements Comprising the Earth's Crust

	Percent		Percent		Percent
Oxygen	49.78	Sodium	2.33	Chlorine	0.21
Silicon	26.08	Potassium	2.28	Carbon	0.19
Aluminum	7.34	Magnesium	2.24	Phosphorus	0.11
Iron	4.11	Hydrogen	0.95	Sulphur	0.11
Calcium	3.19	Titanium	0.37	Total	99.29

SOURCE: Robert Peele, "Mining Engineers' Handbook," John Wiley & Sons, Inc., 1945, p. 1-02.

TABLE 9.2 Classification of Igneous and Volcanic Rocks

Igneous rocks

	Essential minerals	Plagioclase feldspar > ⅔ total feldspar				Little feldspar
		K feldspar	Sodic plagioclase		Calcic plagioclase	Pyroxene and olivine
	Accessory minerals	hornblende, biotite, pyroxene			pyroxene, divine	hornblende, biotite
Homogeneity	Phaneritic	Granodiorite	Qtz diorite	Diorite	Gabbro	Peridotite
Homogeneity	Porphyritic	Graodiorite porphyry / Dacite porphyry	Qtz diorite porphyry	Diorite or andesite porphyry	Gabbro or basalt porphyry	Periodotite porphyry
Homogeneity	Aphanitic	Dacite		Andesite	Basalt	
	Volcanic rocks					
	Aphanitic			Andesite lavas	Basalt lavas	
	Ejecta					

SOURCE: "A Seminar on Current Blasting Practice," The Intercontinental Development Corp., Huntsburg, Ohio, 1974, p. 12.

extrusive rocks is lava flows. Rhyolite and basalt are the most common forms of extrusive rocks.

The grain or crystal difference between intrusive and extrusive rocks results from the cooling time; i.e., the slower the magma cools, the more crystals are allowed to form in the rock. That is why extrusive rocks lack the large crystals and have fine grain and the intrusive rocks have coarser grains with larger crystals.

Minerals in Igneous Rocks There are two types of minerals that form igneous rocks: the felsic (light-colored), which consist of quartz, feldspar, and feldspathoid; and the mafic (darker-colored), which consist of olivine, pyroxene, amphiboles, and micas.

There are subdivisions based on colors, but that is a very rough form of identification, because the weathering of rocks will change the surface coloring. If the rock is to be examined it is best to break a piece to see the internal color.

The best way to examine a rock in the field is by texture, which will tell more of the origin because of the relationship of the type of grain to cooling time, as was previously discussed. Volcanic rock can have a varied texture, which will tell at what stage of crystallization it was when ejected or erupted. If there are both coarse and glossy textures in one rock, then the rock started to crystallize (or cool) before the eruption. Glossy textures indicate that they are volcanic because of the rapid cooling involved, thus preventing the magma from forming crystals.

Sedimentary Rocks

Sedimentary rocks are formed at or near the earth's surface when rocks are eroded either mechanically or chemically or both and the resultant particles are transported and deposited to form sediments. Water is the primary transporter of material that forms sediments; however, the wind and ice also transport sediments.

Chemical erosion is caused by water that has substances or chemicals dissolved in it that attack the rock, causing portions to erode or weaken so that they are eroded mechanically with less effort. Chemical erosion is analogous to the breakdown of non–sulfate-resistant concretes by groundwater high in sulfates.

Mechanical erosion is erosion caused by pressure or forces inflicted on the rock. A tree root's expanding in a crack until it splits a rock is an example of a cause of mechanical erosion.

Weathering is the breakdown of rocks in situ in which the particles are not transported. Weathering may be either mechanical or chemical; however, most climatical weathering is both. Chemical weathering is caused mainly by water, whereas mechanical weathering is from the creeping of rocks and slides or slips. Mechanical weathering can result in a physical shattering of the rock, as when a boulder rolls down a hill and shatters from impact when it reaches the bottom.

Frost is the primary climatical weathering source. As water turns from a liquid to a solid state (freezing), the volume increases about 9%. Therefore, when water lies in the cracks and pores of rock and freezes, the physical expansion of the ice creates pressures on the rock that may reach 2000 lb/in^2 (13,800 kN/m^2). Weathering due to frost is most common at high altitudes and in cold regions. The primary areas for this type of erosion are those where the temperature passes above and below the freezing temperature of water often. This frequent freezing and thawing causes additional strains because of the cyclical or frequent fatigue.

In areas that are exposed to hot sun, the sunlight causes the surface of the rock to expand; then rapid cooling, due to nightfall or rain, can cause surface cracking on the rock. This type of thermal stress is not very effective, and the rock is affected in only a minor way on the surface.

Soft rocks tend to be more affected by rain than harder rocks. Hard rocks, such as quartz, resist weathering. The sandstones and shales are quite resistant to chemical weathering, but the cement that binds the particles together is not.

Erosion caused by water is done generally by rivers, oceans, and groundwater. Rivers erode the vessel or medium in which they flow, i.e., the riverbed and banks. They transport the eroded material, called "load," downstream. There are two components to erosion by rivers: weathering by chemicals in solution (carbonates, sulfates, potassium, sodium, and chlorides of calcium), and mechanical erosion by the particles carried by the river.

The distance the eroded particles are transported depends on their size in relation to the speed and turbulence of the water; i.e., the faster and more turbulent the river, the larger the particles that can be transported. (Particles of rock are named according to their size. See Table 9.3.) The largest particles that the water can move will be rolled or bounced along the river bottom. As the particles decrease in size some of the larger particles will be moved in short, sporadic leaps or jumps; this type of movement is referred to as "saltation." As the slope of the flow line of the river decreases and the velocity of the water decreases, the river drops the load.

The clay (the finest particles) drops when the water becomes still; this is common in floodplain areas along rivers. When the river is rising and the water flows over the surrounding land, it carries particles of clay and silt. When the river crests there is no more flow of water over the land. The clay particles are dropped and can be seen remaining after the water recedes. (See Figure 9.1.)

The erosion of a coastline by the sea is mechanical erosion. The sea batters against the coast, wearing and breaking off pieces of rock, which are, in turn,

TABLE 9.3 Names of Rock Particles According to Size

Particle diameter, mm	Name
1000.0	...
	Boulders
100.0	...
	Cobbles
10.0	...
	Gravel
1.0	...
	Sand
0.1	...
	Silt
0.01	...
	Clay
0.001	...

SOURCE: E. C. Dapples, "Basic Geology for Science and Engineering," Wiley, New York, 1959, p. 20. By permission from John Wiley & Sons, Inc.

Fig. 9.1 The Ohio River floods its banks and leaves sediments as it recedes.

thrown against the coast to create even more erosion. On coastlines with both hard and soft rock, the soft rocks erode first. Many coastal inlets are formed when the soft rock erodes and the hard rock remains. (See Figure 9.2.) Although coastal erosion is mostly mechanical, limestone formations will also be eroded chemically by salt water.

Wind will also erode and transport particles. This erosion is usually restricted to arid and semiarid climates (i.e., areas without much vegetation either because of a lack of rainfall or high elevation) and particularly farmland in such climates. In the desert the wind picks up the sand (which is primarily tough quartz) and bombards rock masses, eroding them. Drought periods make farmland quite susceptible to wind erosion, because most of the natural vegetation has been removed for agriculture and when the crops fail from lack of moisture all the open ground is exposed to movement by the wind. This was witnessed in the Kansas and Oklahoma Dust Bowl of the 1930s.

Ice is also a cause of erosion and a transporter of eroded particles, as can be seen by the study of the glacial period of the earth's history. Glaciers begin as snow, in regions cold enough that the snow does not completely melt during the warm season. As the snow accumulates, the pressure from the increased depth causes sufficient pressure to turn the lower layers of snow to ice. The ice moves slowly down mountainsides, gorging the rock and dragging boulders and rocks it picks up as it moves. The glacier as it moves down into the valleys gouges out a U shape in the sides and bottoms from the original V shape left by erosion by a river or stream. (See Figure 9.3.) The glacier gouges out part of the bottom of the valley deeper than the surrounding area; this area may eventually be filled with water to form a lake.

The rocks and material carried by the glacier are dropped as the glacier begins to melt. This process has caused the finding of boulders that are not of minerals indigenous to the area in which they are found.

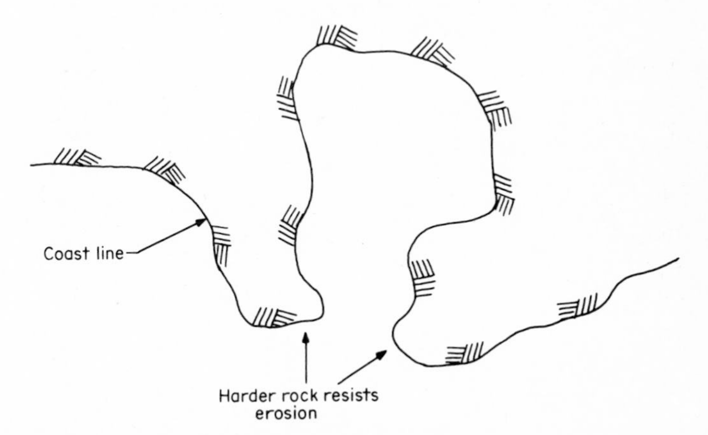

Fig. 9.2 Natural harbors formed by erosion.

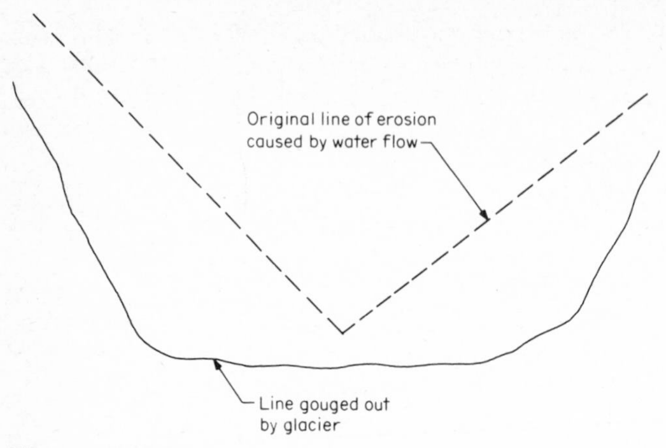

Fig. 9.3 How a glacier carves out a valley.

Diagenesis The process by which loose, unconsolidated particles are changed into cohesive sedimentary rocks by low temperature and pressure near the surface of the earth is called "diagenesis." This process requires no movement of the earth's crust, and the most common method is by the compaction and cementation created by the weight of the additional sediments on top; i.e., as sediments are added, the sediments on the bottom are subjected to the weight of the higher sediments. As the material is compacted the water is forced, or squeezed, out. The soluble and unstable particles or minerals dissolve and enter the pores of the sediment, forming cement; or the cement may be added to the sediment by groundwater. The common cements for sedimentary rock are alute, silica, and iron hydrates. The final stage in the sedimentation process, where loose particles are cemented, forming into rock mass, is called "lithification." If the compaction pressure is extremely

TABLE 9.4 Classification of Sedimentary Rocks

Origin	Texture	Particle size or composition	Rock name
Detrital	Clastic	Mixed-granule or larger Sand Silt and clay	Conglomerate Sandstone Shale or mudstone
Chemical Inorganic Biochemical	Clastic and nonclastic	Calcite ($CaCO_3$) Dolomite ($CaMg(CO_3)_2$) Halite (NaCl) Gypsum ($CaSO_4.2H_2O$) Calcite ($CaCO_3$) Plant remains	Limestone Dolomite Salt Gypsum Limestone Coal

SOURCE: L. D. Leet and Sheldon Judson, "Physical Geology," 2d ed., Prentice-Hall, New York, 1958, p. 112. By permission from Prentice-Hall, Inc., Englewood Cliffs, N.J.

Fig. 9.4 Sedimentary rock formation along a highway. (The black layer is coal.)

high the particles may become recrystallized. This recrystallizing of the sediment is a metamorphic process.

Types of Sedimentary Rocks There are two major types of sedimentary rocks: clastic and nonclastic. Clastic rocks are those that are formed by the cementing of particles from other rocks and are classified by base or grain size. Nonclastic rocks are formed by chemical or organic processes. (See Table 9.4. Also refer to Table 9.1 for the mineral composition of sedimentary rocks.)

The most common types of sedimentary rocks are sandstone, shale, conglomerates, breccia, limestone, dolomite, and bituminous coal. (See Figure 9.4.)

Sandstone is made up of sand and cement. It is porous and permeable to fluids. Sandstones are clastic, many of the particles being quartz. Sandstone is fairly soft to drill but is quite abrasive to drill bits. Siltstone is quite similar to sandstone, but the grains are much finer. Shale is formed from the compaction of mud and clay. It has a hardness of 2 and is quite soft to drill. (Table 9.5 explains the scale of hardness.)

Conglomerates are sedimentary rocks that are formed from gravel and are often referred to as "pudding stone." Conglomerates have a visual likeness to concrete and are as unpredictable as concrete to blast. Breccia is similar to conglomerates except that the particles are angular rather than rounded.

Limestone is almost entirely calcium carbonate, with some clay and sand. Limestone is formed on the sea bottom, and the proximity to the shore deter-

TABLE 9.5 Mohs' Scale of Hardness

Hardness	Defining mineral	Scratch test
10	Diamond	...
9	Corundum	...
		Tungsten carbide
8	Topaz	...
		Hardened steel
7	Quartz	...
6	Feldspar	Steel
5	Apatite	Glass
4	Fluorspar	...
		Brass pin
3	Calcite	...
		Fingernail
2	Gypsum	...
1	Talc	...

mines the quantity of clay found in it; i.e., the closer to shore, the more clay. It is 3 to 4 in hardness; however, its drilling hardness can vary greatly, depending on how the particles are cemented.

Dolomite is similar to limestone except that it contains magnesium carbonate. Often a chemical analysis is necessary to distinguish between it and limestone.

Bituminous coal is nonclastic soft coal that is primarily carbon from decayed vegetable matter.

TABLE 9.6 Common Metamorphic Rocks

Name	Rock texture	Commonly formed by metamorphism of
	Unfoliated	
Quartzite	Granular (breaks through grains)	Sandstone
Marble	Granular	Limestone, dolomite
Hornfels	Dense	Fine-grained rocks
	Foliated	
Slate	Fine-grained	Shale, mudstone
Phyllite	Fine-grained	Shale, mudstone
Schist	Fine-grained	Shale, mudstone, andesite, basalt
Gneiss	Coarse-grained	Granite

SOURCE: L. D. Leet and Sheldon Judson, "Physical Geology," 1st ed., Prentice-Hall, New York, 1954, p. 342. By permission from Prentice-Hall, Inc., Englewood Cliffs, N.J.

Metamorphic Rocks

Metamorphic rocks are rocks that have undergone a change in mineral content, texture, or both from their igneous or sedimentary predecessors. Metamorphic rocks are formed by high temperature and great pressure from deep within the earth at temperatures ranging from 400 to 1800°F (204 to

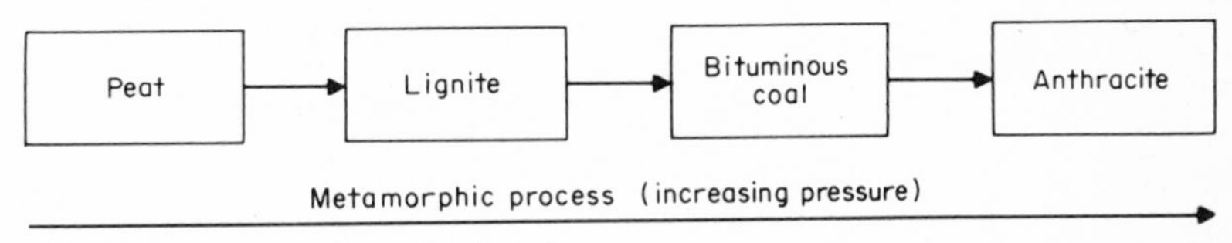

Fig. 9.5 The formation of anthracite.

982°C). The pressure is from the weight of the overlying rock or from movements in the earth's crust. The pressure and heat in metamorphic development cause the atoms of the minerals to rearrange, creating new materials. Also the reduction in volume creates denser materials. For example, garnet is produced during the metamorphic process acting upon granite. Table 9.6 lists several metamorphic rocks and the rocks from which they are formed.

The most common or easily traceable process is that of anthracite. Anthracite is the metamorphic counterpart of bituminous coal. Figure 9.5 shows the development stages of anthracite.

DEFORMATION OF ROCK

Rock can be considered either elastic or plastic. Elastic rock will deform up to the point of rupture and will fracture from shear, producing joints and faults that relieve some of the stresses in the rocks. Plastic rock is weak and will deform easily. It is abundant in clay-type material and will usually contain few fractures. If fractures do develop they are the result of tensile stresses. Very stiff plastic rock may fracture in either tension or shear.

Folds

Folds are wrinkles or warps where regular patterns of crests and troughs can be observed in some rock formation. Folds are generally more often observed in layered rocks, where the differences in color or texture of the rock make it easier to see the folds. When a fold is concave, i.e., forming a trough, it is called a "syncline," and when the fold is convex, forming a crest, it is called an "anticline." (See Figure 9.6.)

There are two fundamental types of folds: flexural slip folds and slip folds. Flexural folds are created when adjacent layers slip past each other with the layers retaining their original thickness. Slip folds, also known as "shear folds," are found primarily in sedimentary or metamorphic rocks under ex-

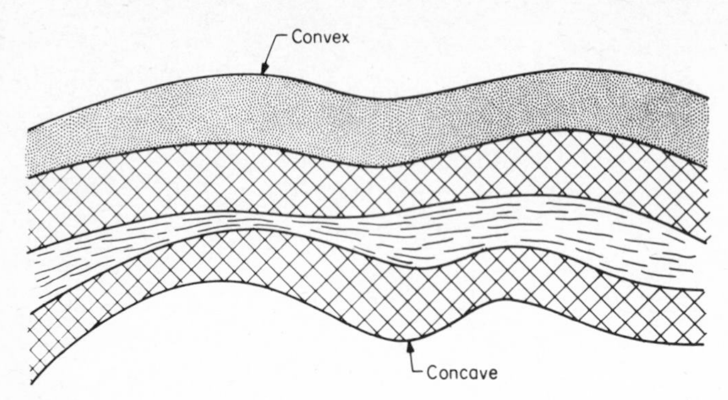

Fig. 9.6 A folded belt.

treme confining pressure; the folding is caused by slipping along closely spaced slip planes.

Faults

A fault is the result of an accumulation of stresses in the rock to the point of rupture and a displacement along the area of rupture. If groundwater is near a fault, the temperature of the water is increased and chemical reactions are prompted, causing changes in the rock. This type of change or alteration about faults is called "argillization." Argillization produces clay-type minerals at the cost of feldspars, creating a high degree of plasticity in materials in the fault area. Faults are generally the shear planes along which the movement or concentration of forces takes place.

Types of Faults There are three types of faults: normal, reverse, and wrench faults (see Figure 9.7). Normal faults are faults in which the block lying above an inclined fault surface moves downward relative to the other block; they are due to tensile stresses. Reverse, or thrust, faults are faults in which the block above the incline moves upward relative to the other block. Mylonites—zones of soft, pulverized rock that are most common with granite—are often formed in areas that have been subjected to reverse faults. This type of rock can be an area prone to a lot of overbreak when blasted. Wrench faults are the result of shear stresses along a vertical plane that cause a horizontal movement of the blocks relative to each other.

Joints

A joint is a fracture in rock along which there has been little or no visible differential displacement in a direction parallel to the fracture plane. "Joint" is the name given to the cracks seen in rock; although they may appear to be haphazard, with study, definite patterns can be found, showing that the cracks were caused by the relieving of stresses. In igneous rocks the cracks are generally from cooling, whereas with sedimentary rocks the cracks, or joints, are caused by desiccation, or drying. Joints can have an attitude; i.e., joints may be vertical, horizontal, or at an incline. It is the absence of movement that distinguishes a joint from a fault.

Joints can affect the blast design in many ways. The joints can hurt the breakage by allowing the premature release of gases, prohibiting the compression wave from reaching the face because of refraction and reflection from excessive jointing, and can allow the rock to fragment only at the joints,

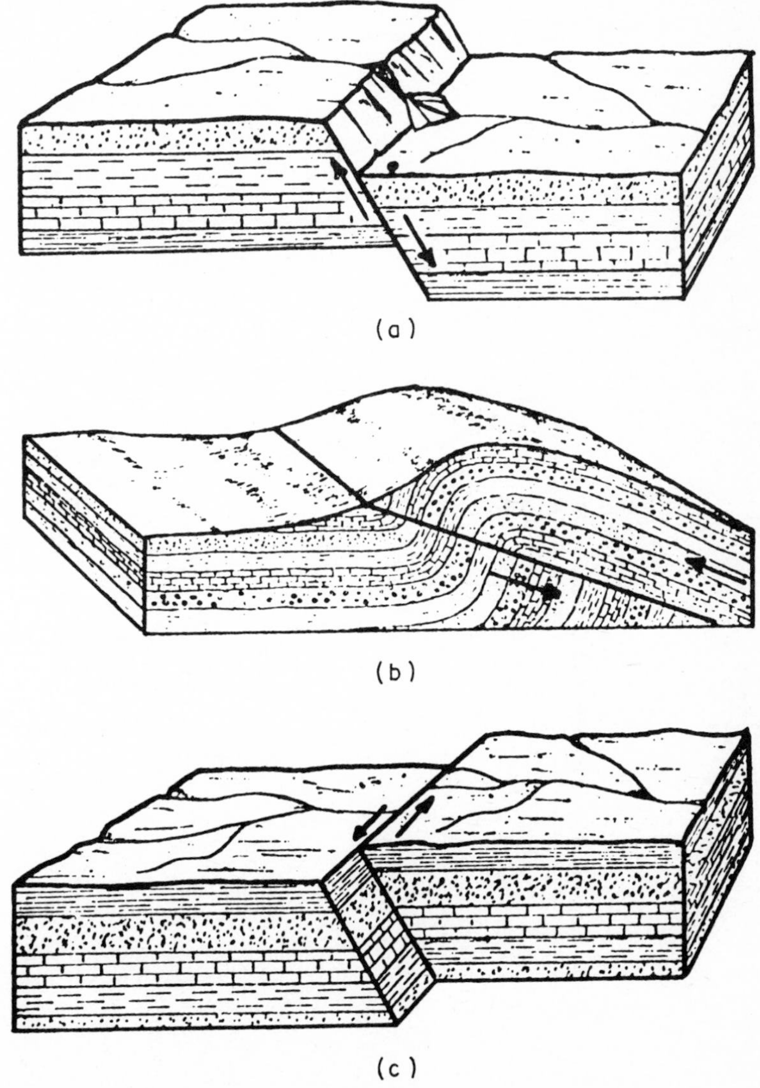

Fig. 9.7 Faults: (*a*) normal fault; (*b*) reverse fault; (*c*) wrench fault.

TABLE 9.7 Engineering Properties for Some Selected Rock Types

Rock type	Location	Tensile strength, lb/in^2	Compressive strength, lb/in^2	Modulus of elasticity, $lb/in^2 \times 10^6$	Longitudinal velocity, ft/s	Poisson's ratio	Specific gravity
Amphibolite	India	—	61,400	—	19,000	—	3.12
Basalt	U.S.	1,552	21,630	9.04	17,150	0.275	2.88
Dolomite	U.S.	366	7,943	4.10	13,200	0.325	2.54
Gneiss	U.S.	1,985	32,551	11.75	18,800	0.217	2.80
Granite	U.S.	1,308	27,020	6.27	15,888	0.327	2.70
Limestone	W. Virginia	750	23,000	8.00	16,400	0.250	2.68
Marble	U.S.	2,206	36,400	15.39	21,992	0.284	3.04
Sandstone	W. Virginia	200	19,400	1.00	12,900	—	2.51
Sandstone	U.S.	40	1,540	0.86	6,870	0.309	1.87
Schist	U.S.	1,330	24,010	11.15	17,980	0.203	2.85
Serpentine	U.S.	740	16,351	7.72	17,633	0.333	2.75
Slate	U.S.	927	12,390	9.56	16,950	0.167	2.64
Taconite	Minnesota	2,474	36,401	13.48	20,140	0.249	2.95

SOURCE: "A Seminar on Current Blasting Practice," The Intercontinental Development Corporation, Huntsburg, Ohio, 1974, p. 152.

prohibiting small enough fragments (i.e., producing many boulders). Sometimes the rock may be soft enough and contain so many cracks that it may be possible to remove it with ease with a ripper.

When highly jointed rock is blasted it may be necessary to stay with cartridged explosives because of the cracks: when AN/FO is poured into holes that contain many cracks, the cracks can sometimes be large enough to catch some of the AN/FO. Not only may this waste explosives, it may also be dangerous. If there is a large crack or hole that fills with explosive, one ends up with a concentrated charge that may cause excessive flyrock.

ENGINEERING PROPERTIES

The engineering properties of rock that are considered in blast design are (1) the density, (2) the strength (primary tensile and shear), (3) the longitudinal wave velocity, (4) the modulus of elasticity, and (5) Poisson's ratio. (See Table 9.7 for values of these for several selected rocks.)

Density

The density of rock can be expressed by specific gravity or unit weight. The specific gravity of a rock is the rock's weight relative to the weight of water for the same volume. The specific gravity of granite is 2.7; that is, granite will weigh 2.7 times its volume in water. The density is the unit weight of the material. The easiest way to calculate the density is to multiply the specific gravity by the unit weight of water. The density is generally expressed in pounds per cubic foot.

Density of granite:

$$\begin{aligned} \text{Specific gravity of granite} &= 2.7 \\ \text{Density of water} &= 62.4\ \text{lb/ft}^3 \\ 2.7(62.4\ \text{lb/ft}^3) &= 168.5\ \text{lb/ft}^3 \end{aligned}$$

Strength

By "strength" we mean ultimate strength, i.e., the maximum unit stress attained before rupture. Before the ultimate strength is reached the rock is considered plastic. That is, when the stress is relieved the rock will regain its shape. When the ultimate strength is passed the rock will crack. Table 9.8 gives values for the strength of selected rocks.

Longitudinal-Wave Velocity and Modulus of Elasticity

The longitudinal-wave velocity is the velocity at which the rock will transmit, or propagate, the compression waves. The denser and more homogeneous the rock, the more efficiently the waves are propagated. The detonating velocity of the explosive required is determined by the wave-propagating characteristics of the rock. The higher the longitudinal velocity of the rock,

TABLE 9.8 Strength of Rocks*

Rock	Modulus of rupture, lb/in²	Compressive strength, lb/in²	Tensile breaking strain,† in/in × 10^{-6}
Granite	2700–3900	28,400–39,500	380
Basalt	2500–6600	26,600–52,000	
Sandstone	400–3600	6,100–34,100	560
Shale	2200–5000	15,600–45,800	
Limestone	400–3800	9,700–37,600	310
Marble	1700–3300	18,000–33,000	
Gneiss	1200–3100	22,200–36,400	
Amphibolite‡	4200–7400	30,400–74,900	
Greenschist§	3200–6700	17,700–45,500	

* S. L. Windes, Physical Properties of Mine Rock, *U.S. Bur. of Mines Rept. of Investigations* 4459, Government Printing Office, Washington, 1949.
† W. I. Duvall and T. C. Atchison, Rock Breakage by Explosives, *U.S. Bur. of Mines Rept. of Investigations* 5356, Government Printing Office, Washington, 1957.
‡ Amphibolite is a faintly foliated metamorphic rock composed mainly of hornblende and plagioclase feldspars, developed during the metamorphism of simatic rocks.
§ Greenschist is a schist characterized by green color imparted by the mineral chlorite. It is formed by the metamorphism of simatic rocks.

SOURCE: L. D. Leet, "Vibrations from Blasting Rock," Harvard University Press, Cambridge, 1960, p. 42.

TABLE 9.9 Longitudinal-Wave Speeds and Characteristic Impedances for Certain Rocks

Rock	Longitudinal-wave speed, ft/s	Characteristic impedance, lb-s/in³
Granite	18,200	54
Marlstone*	11,500	27
Sandstone	10,600	26
Chalk†	9,100	22
Shale	6,400	15

* Marlstone is a hardened mixture of clay materials and calcium carbonate, normally containing 25 to 75% of clay. It is a type of limestone.
† Chalk is a very soft limestone.

SOURCE: L. D. Leet, "Vibrations From Blasting Rock," Harvard University Press, Cambridge, 1960, p. 42.

the higher the required velocity. The modulus of elasticity E is the measurement of stiffness of the rock; i.e., it is the measurement of the rock's ability to withstand or resist deformation. The higher the modulus of elasticity, the harder the rock is to break. (See Table 9.9.)

Poisson's Ratio

Poisson's ratio is the relationship of lateral deformation to longitudinal deformation, i.e., the degree to which, as a material stretches or compresses under a load, the cross-sectional area perpendicular to the stress decreases or increases. Poisson's ratio is of no value in determining the blasting characteristics of rock except that the lower the ratio is, the more prone the rock is to presplitting.

We have shown in this chapter that the character of the rock, its origin, joints, and engineering properties have a definite effect on the blasting methods and materials used.

TEN

VIBRATIONS AND HIGH-LIABILITY BLASTING

BLASTING CLAIMS

One of the largest single technical problems that faces a blasting contractor is blasting claims. Although this at first seems more a legal problem, its solution is of a technical nature. It is important that all of those engaged in or having responsibility for blasting operations have a basic knowledge or understanding of the causes and cures of blasting claims.

There are four causes of blasting claims, or damage: air concussion, airborne shock waves, earth vibration, and flying debris (generally referred to as "flyrock").

Air Concussion

Air concussion, or air blast, is a pressure wave traveling through the air; it is generally not a problem in construction blasting. The type of damage created by air concussion is broken windows. However, it must be realized that a properly set window glass can generally tolerate pressures of up to 2 lb/in^2 (13.8 kN/m^2), whereas a wind of 100 mi/h (161 km/h) produces a pressure of only 0.35 lb/in^2 (2.4 kN/m^2).

Air concussion is caused by the movement of a pressure wave generally caused by one or more of three things: a direct surface energy release, a release of inadequately confined gases, and a shock wave from a large free face (see Figure 10.1). Direct surface energy release is caused by detonation of an explosion on the surface. Detonating cord, when exploded, would be a source of this energy release. Premature release of gases into the atmosphere when a column of explosives is detonated is due to poor stemming or mud seams or both. Finally, when there is a very large free face, as is often the case in quarry blasting, the ground movement can produce an airborne pressure wave.

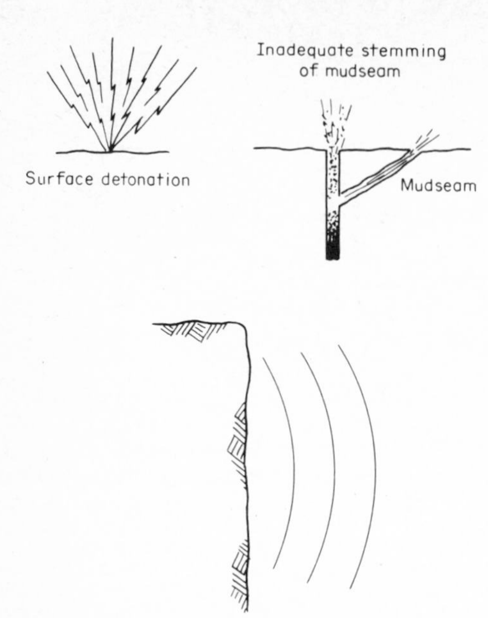

Fig. 10.1 Three causes of air blast.

Air concussion is also referred to as "overpressure," i.e., the air pressure over and above normal air pressure. The difference between noise and concussion is the frequency: air blast frequencies below 20 hertz (Hz), or cycles per second, are concussions, because they are inaudible to human ears; whereas any air blast above 20 Hz is noise, because it is audible.

Overpressure can be measured in two ways: in pounds of pressure per square inch (lb/in²) or in decibels (dB). Decibels are a measurement or expression of the relative difference of power of sound waves. Pounds per square inch and decibels can be made relative by the following equation:

$$\text{overpressure (dB)} = 20 \log \frac{p}{p_0}$$

where overpressure (dB) = overpressure in decibels

log = the common logarithm

p = overpressure in lb/in²

$p_0 = 3 \times 10^{-9}$ lb/in²

For example, if we have an overpressure of 0.5 lb/in², the figure in decibels is

$$\text{overpressure (dB)} = 20 \log \frac{0.5}{3 \times 10^{-9}}$$
$$= 164.4 \text{ dB}$$

Figure 10.2 is a chart of the effects of several values of overpressure given as numerical ratios and in decibels.

Airborne Shock Wave

Unlike air concussion, airborne shock waves generally travel considerable distances from the site of the blast at the speed of sound. These shock waves are caused by an explosion on the surface that produces a pressure wave moving at supersonic speed for 20 to 50 diameters of the explosives before slowing to the speed of sound. Its total distance and intensity are governed by the terrain, atmospheric conditions, and obstacles. Generally these pressure waves achieve the greatest distance traveling upward until they are dissipated in the atmosphere. Of course, terrain and atmospheric conditions can affect the behavior of these pressure waves. Such was the case in San Antonio, Texas, when 112,000 lb (50,800 kg) of explosives was accidentally detonated. Most of the damage was 7 to 10 mi (11.2 to 16 km) from the explosion. In this case it was caused by an inversion layer at an altitude of 5000 ft (1524 m). As with air concussion, glass breakage is the first damage to occur; one will find no damage without glass breakage.

Vibrations

The most common types of blasting claim are those due to earth movement, or vibrations. Although noise is often what produces the complaints, vibration generally is the cause of damage not caused by flying debris.

When using explosives to break rock there are a number of effects: noise (which can produce complaints but generally not damage), total displacement in the immediate area around the explosive, plastic deformation, and elastic motion. Total displacement in the immediate area around the explosive is what one wishes to accomplish: it is intended that the affected rock

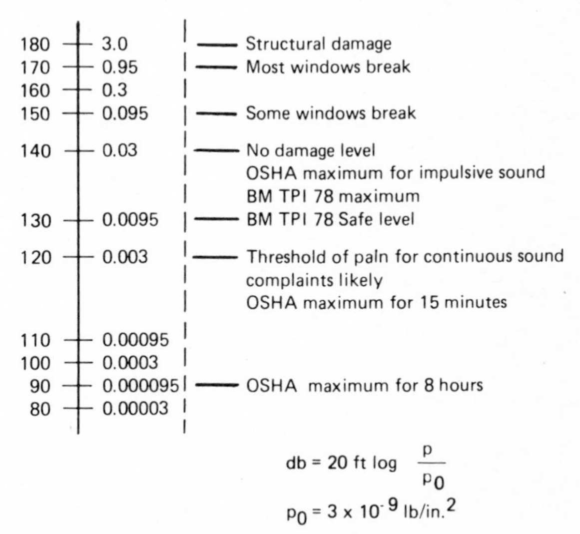

Fig. 10.2 Air blast overpressure scale. (*Atlas Powder Co.*)

change position or location and a permanent differential deformation occur changing the size and shape of the rock.

The plastic zone is the area just a few feet beyond the planned displacement. Researchers still do not understand exactly what happens in this region; however, it is of no consequence to the usual blasting claims.

Elastic motion is that back-and-forth vibration motion in which frequency and amplitude are predictable from the variables, the terrain, and the amount of explosives. This is the cause of the majority of blasting claims. However, as one investigates this further, it will be obvious that these claims are infrequently accurate.

When an explosive that is buried in the ground is detonated most of its energy is spent shattering the rock or other materials around it. However, since an explosion is an imperfect use of energy, there is a loss of some energy transmitted through the earth in the form of waves or vibration. Some waves will still escape in the form of noise or concussion. There are two principal classes of waves: one that travels through the interior of the earth, and another that travels on the surface. Approximately 95% of the earthborne energy reaches the surface a few feet from the total displacement area and acts as surface waves. These surface waves are analogous to the ripples that develop when a stone is cast into a pond.

Amplitude The two terms generally used to describe a wave are "amplitude," which is half the distance between the crest and the trough (see Figure 10.3), and "frequency," which is the number of complete waves that pass a given point per unit of time. An easier way to visualize amplitude and frequency is to picture a small boat on the water. As it moves up and down on the waves, the highest point the boat reaches is the crest and the lowest it reaches is the trough. Half the distance between the crest and trough is the amplitude. The number of times the boat moves up and down in one second is called the frequency. In a structure, we find that the amplitude (displacement) is affected by four factors: the amount (weight) of explosives used per blast, the geologic material that the waves pass through, the distance between the structure and the blast, and the type of material under the structure.

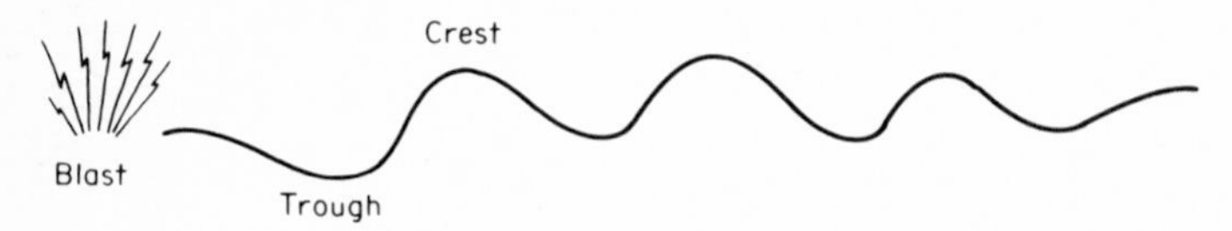

Fig. 10.3 Vibration wave.

Vibration causes damage by differential displacement. As waves pass under a structure, they will lift the structure up and down, from side to side, and back and forth. However, if this movement of the structure could be in

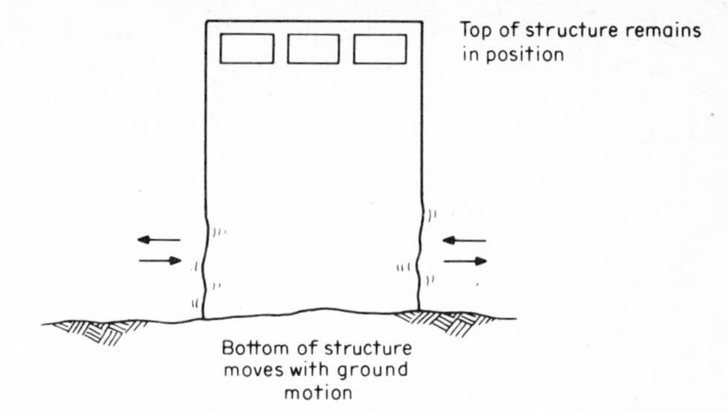

Fig. 10.4 How differential displacement causes structural damage.

its entirety, there would be no damage. It is the differential movement that actually causes the damage. Theoretically you could turn the structure upside down and set it back down in the original position without damage, provided the structure moved monolithically. However, in actual practice, structures subjected to ground movement generally resist the movement, creating differential loading and therefore stress. Generally, while the lower portion is in motion due to vibration, the top of the structure is in its original position at rest. (See Figure 10.4.) Stresses of this nature generally produce scissor cracks 45° to the horizontal and at 90° to each other. (See Figure 10.5.)

Vibration Damage Vibration damage usually first appears as extensions of old cracks. Because plaster is usually the weakest material in the building it is the first material to form new cracks.

There are common misconceptions about blasting and the damages caused by vibration. Usually people believe the louder the noise is, the greater the damage. There is not necessarily a relationship between the two. The main reason for people's concern about blasting damage is that the human body can readily feel the effects of vibration. Some people have been tested and have been found able to detect vibration at a level one one-

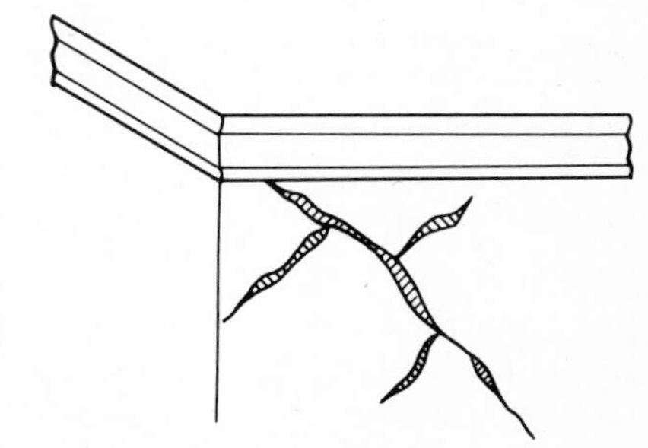

Fig. 10.5 Typical plaster cracks. (Note how they form at 90° to each other.)

hundreth of the level necessary to damage structures. Most of the time, normal activities will produce more vibration than blasting. (See Table 10.1.)

Bureau of Mines Tables: One way to avoid claims is to determine the amount of explosives that can be used per blast without causing damage. The best way to calculate this allowable charge is by using the tables developed by the U.S. Bureau of Mines in Bulletin 442. Table 10.2 gives the displacement in inches for various weights of explosives. Using the distance from the blast and the amount of explosives, you can find the figures on the table. For example, using 50 lb (22.7 kg) at 400 ft (122 m) we find 0.0055 in. However, these displacements are shown for average overburden (less than 50 ft). If displacements are to be computed for abnormal overburden (more than 50 ft) or abnormally responsive overburden such as sand, gravel, or loam, the displacements from the table should be multiplied by 3. If displacements are computed for outcrops of rock the displacement from the table should be multiplied by 0.10. For the example of 50 lb at 400 ft, if we

TABLE 10.1 Vibrations from Normal Activities

	Particle velocity in room, in/s*			Particle velocity in adjacent room, in/s		
Activity	Radial	Vertical	Transverse	Radial	Vertical	Transverse
Walking	0.00914	0.187	0.372	0.00129	—	0.00102
	—	0.0578	0.0155	0.00167	0.0281	0.00227
	—	0.00770	0.00210	0.00229	0.0626	0.00462
	0.0600	0.120	0.0300	—	—	—
	0.0100	0.0600	0.007	—	—	—
	0.00600	0.0110	0.00400	—	—	—
	0.00800	0.0200	0.00700	—	—	—
Door closing	0.0110	0.0558	0.0149	0.00170	—	0.00153
	—	0.0150	0.00500	0.0125	0.0970	0.00963
	0.008	0.0100	0.00800	—	—	—
Jumping	0.0524	4.03	1.05	0.120	0.219	0.551
	0.120	0.219	0.551	0.0153	0.0239	0.0101
	1.00	2.500	1.70	0.00450	0.0100	0.0045
	0.500	5.00	1.10	—	—	—
Automatic washer	0.00340	0.00400	0.00340	—	—	—
Clothes dryer	0.00500	0.00500	0.00500	—	—	—
Heel drops	0.0100	0.0100	0.0100	—	—	—
	0.0800	0.600	0.0300	—	—	—
	0.0200	0.200	0.0200	0.006	0.0100	0.006
	0.900	3.500	0.400	—	—	—
	0.0500	0.450	0.0700	0.009	0.014	0.008
	0.0100	0.200	0.00900	—	—	—

* 1 in/s = 0.0254 m/s.

SOURCE: Blasting Vibrations and Their Effects on Structures, *Bureau of Mines Bulletin* 656, 1971, p. 21.

TABLE 10.2 Displacement for Various Weights of Explosives, in

Weight of explosives, lb	Distance from blast, ft								
	100	200	300	400	500	600	700	800	900
10	0.0029	0.0025	0.0022	0.0019	0.0016	0.0014	0.0013	0.0011	0.0010
20	0.0045	0.0039	0.0034	0.0030	0.0026	0.0023	0.0020	0.0017	0.0015
30	0.0059	0.0052	0.0045	0.0049	0.0034	0.0030	0.0026	0.0022	0.0020
40	0.0072	0.0063	0.0054	0.0047	0.0041	0.0036	0.0032	0.0027	0.0024
50	0.0084	0.0073	0.0063	0.0055	0.0048	0.0042	0.0027	0.0032	0.0028
60	0.0059	0.0082	0.0072	0.0063	0.0054	0.0047	0.0042	0.0036	0.0031
70	0.010	0.0091	0.0079	0.0069	0.0060	0.0052	0.0047	0.0039	0.0035
80	0.011	0.0099	0.0086	0.0075	0.0065	0.0057	0.0051	0.0043	0.0038
90	0.012	0.011	0.0093	0.0081	0.0070	0.0061	0.0055	0.0046	0.0041
100	0.013	0.012	0.010	0.0087	0.0076	0.0066	0.0059	0.0050	0.0044
200	0.021	0.018	0.016	0.014	0.012	0.010	0.0094	0.0079	0.0069
300	0.0028	0.024	0.021	0.018	0.016	0.014	0.012	0.010	0.0091
400	0.033	0.029	0.025	0.022	0.019	0.017	0.015	0.013	0.011
500	0.039	0.034	0.029	0.026	0.022	0.019	0.017	0.015	0.013
600	0.044	0.038	0.033	0.029	0.025	0.022	0.019	0.016	0.014
700	0.049	0.042	0.037	0.032	0.028	0.024	0.022	0.018	0.016
800	0.053	0.046	0.040	0.035	0.030	0.026	0.024	0.020	0.017
900	0.057	0.050	0.043	0.038	0.033	0.029	0.026	0.022	0.019
1,000	—	—	—	—	0.035	0.031	0.027	0.023	0.020
2,000	—	—	—	—	0.056	0.049	0.044	0.037	0.032
3,000	—	—	—	—	0.073	0.064	0.057	0.048	0.042
4,000	—	—	—	—	0.089	0.078	0.069	0.059	0.051
5,000	—	—	—	—	0.10	0.090	0.080	0.068	0.059
6,000	—	—	—	—	0.12	0.10	0.090	0.076	0.067
7,000	—	—	—	—	0.13	0.11	0.10	0.085	0.074
8,000	—	—	—	—	0.14	0.12	0.11	0.93	0.081
9,000	—	—	—	—	0.15	0.13	0.12	0.10	0.088
10,000	—	—	—	—	0.16	0.14	0.13	0.11	0.094

Weight of explosives, lb	Distance from blast, ft					
	1000	2000	3000	4000	5000	6000
100	0.0038	0.0011	0.0004	0.0003	0.0002	0.0002
200	0.0060	0.0017	0.0007	0.0004	0.0003	0.0003
300	0.0079	0.0022	0.0009	0.0005	0.0004	0.0004
400	0.0096	0.0027	0.0011	0.0006	0.0005	0.0005
500	0.011	0.0032	0.0013	0.0008	0.0006	0.0006
600	0.013	0.0036	0.0014	0.0009	0.0007	0.0007
700	0.014	0.0039	0.0016	0.0009	0.0008	0.0008
800	0.015	0.0043	0.0017	0.0010	0.0009	0.0009
900	0.016	0.0047	0.0019	0.0011	0.0009	0.0009
1,000	0.018	0.0050	0.0020	0.0012	0.0010	0.0010
2,000	0.028	0.0080	0.0032	0.0019	0.0016	0.0016
3,000	0.037	0.010	0.0042	0.0025	0.0021	0.0021
4,000	0.045	0.013	0.0051	0.0030	0.0025	0.0025
5,000	0.052	0.015	0.0058	0.0035	0.0029	0.0029
6,000	0.058	0.016	0.0066	0.0040	0.0033	0.0033
7,000	0.065	0.018	0.0073	0.0044	0.0036	0.0036
8,000	0.071	0.020	0.0080	0.0048	0.0040	0.0040
9,000	0.076	0.022	0.0086	0.0052	0.0043	0.0043
10,000	0.082	0.023	0.0093	0.0056	0.0046	0.0046

SOURCE: Seismic Effects of Quarry Blasting, *Bureau of Mines Bulletin* 442, 1942.

have very responsive overburden, we multiply 0.0055 in by 3. Taking this figure (0.0165 in) and looking at Table 10.3 gives the acceleration and its safe limits. Using the ground conditions of Table 12.2, we have for average overburden 15 Hz, for abnormal overburden 5 Hz, and for rock outcrops 20 to 80 Hz. With the same overburden as in Table 10.2, the value is 5 Hz. Interpolating between 4 and 6 Hz on Table 12.3 and comparing the displacement we find that it is a safe level.

The Scaled Distance Formula: Another method for determining the safe limits for ground vibrations is the scaled distance formula.

$$W = \left(\frac{D}{D_s}\right)^2$$

where $D_s = 60$, D = distance, and W = maximum explosive weight, lb. Using the previous example, 50 lb at 400 ft, we find $W = (400/60)^2$. This would indicate that the blaster should not use that much explosive. Using the tables we found that this amount is acceptable, but the scaled distance formula does not permit it. Obviously the tables are more accurate; however,

TABLE 10.3 Acceleration

Displacement, in	Frequency, Hz 2	4	6	8	10	15	20
0.24	0.1	0.38	0.86	1.5	2.4	5.4	9.6
0.22	0.09	0.35	0.79	1.4	2.4	5.0	8.8
0.20	0.080	0.32	0.72	1.3	2.0	4.5	8.0
0.18	0.072	0.29	0.65	1.2	1.8	4.1	7.2
0.16	0.064	0.26	0.58	1.0	1.6	3.6	6.4
0.14	0.056	0.22	0.50	0.90	1.4	3.2	5.6
0.12	0.048	0.19	0.43	0.77	1.2	2.7	4.8
0.10	0.040	0.16	0.36	0.64	1.0	2.2	4.0
0.08	0.032	0.13	0.29	0.51	0.8	1.8	3.2
0.06	0.024	0.10	0.22	0.38	0.6	1.3	2.4
0.04	0.016	0.06	0.14	0.26	0.4	0.9	1.6
0.02	0.008	0.03	0.07	0.13	0.2	0.4	0.8
0.01	0.004	0.016	0.036	0.064	0.1	0.2	0.4
0.008	0.0032	0.013	0.029	0.051	0.08	0.2	0.3
0.006	0.0024	0.010	0.022	0.038	0.06	0.1	0.2
0.004	0.0016	0.006	0.014	0.026	0.04	0.09	0.2
0.002	0.0008	0.003	0.007	0.013	0.02	0.04	0.08
0.001	0.0004	0.0016	0.0036	0.006	0.01	0.02	0.04
0.0008	0.0003	0.0013	0.0029	0.005	0.008	0.02	0.03
0.0006	0.0002	0.0010	0.0022	0.004	0.006	0.01	0.02
0.0004	0.0002	0.0006	0.0014	0.0026	0.004	0.01	0.016
0.0002	0.0001	0.0003	0.0007	0.0013	0.002	0.004	0.008
0.0001	0.0000	0.0002	0.0004	0.0006	0.001	0.002	0.004

Safe / Caution / Damage

SOURCE: Seismic Effects of Quarry Blasting, *Bureau of Mines Bulletin* 442, 1942.

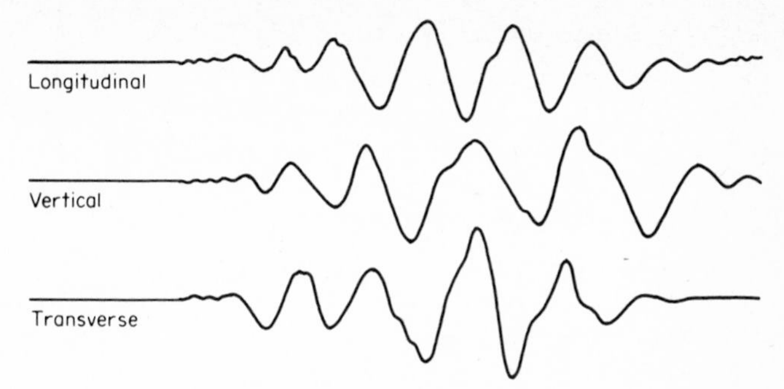

Fig. 10.6 A typical seismogram. (*Atlas Powder Co.*)

the scaled distance formula is a good, expedient method in the field for avoiding vibration problems.

The Seismograph: The best method for monitoring the vibrations due to blasting is with a seismograph. The seismograph will record the vibrations on a seismogram. These recordings will measure the displacement velocity, acceleration, and frequency.

The displacement is the distance, generally in a fraction of an inch, that oscillating particles move. Velocity is the speed with which the oscillating particles move, in inches per second. The velocity will vary; therefore, the peak velocity is recorded. Acceleration is the rate at which the particles change velocity. This is referred to in either feet per second or g's, g being the acceleration due to earth's gravity (32.2 ft/s^2). Frequency is the number of oscillations that the particle undergoes per second. In the example of the small boat riding the waves, if the boat were to ride over two waves in 1 s, it would have a frequency of 2 Hz. Vibration damage will usually occur in frequencies of 3 to 100 Hz.

The typical seismogram (see Figure 10.6) has three traces recorded on it: longitudinal, vertical, and transverse. These traces measure, respectively, particle movement back and forth in the direction that the wave is traveling, up-and-down particle movement, and movement left and right, or perpendicular to the direction the wave is traveling.

At one time the U.S. Bureau of Mines determined that, although it generally requires particle velocities in excess of 4.2 in/s to cause structural damage, 2 in/s was the maximum safe range. However, currently the maximum particle velocity permitted by the Office of Surface Mining is 1 in/s.

Reducing Vibration Blast design is the key to reducing vibrations from blasting: it makes it possible to reduce the amount of explosive that is detonated at one time. Depending on the situation there are several methods for reducing the size of an individual blast. One or several of the methods may be incorporated into one blast.

Delay Blasting: First is the use of delay blasting caps. If there is 0.008 sec (8 ms) between blasts, then for vibration purposes they are considered independent blasts. Therefore, with delay blasting caps the amount of explosives used for one blast is reduced. In the example in Figure 10.7, if there is a shot that requires 12 holes with 10 lb (4.54 kg) of explosives in each, that will produce a one-time, or instantaneous, blast of 120 lb (54.43 kg). However, if the blast could be designed to use seven delays then the maximum weight of explosives detonating at one instant would be reduced to 20 lb (9.08 kg).

Reducing Hole Diameter: Another method for reducing the weight of explosives to be detonated at one instant is to reduce the diameter of the borehole. This causes a reduction in the burden and spacing. However, this decrease in the diameter of the borehole reduces the loading density and thus decreases the amount of explosive detonated at one instant. This reduction has to be accompanied by detonating the same number of holes at one time as planned originally or utilizing delay blasting caps. It should be noted that reducing the hole diameter and keeping the same volume of rock to be blasted will accomplish nothing with regard to vibration control. The purpose of reducing the hole size is to create a reduction in explosives detonated at one instant. However, if the total blast size remains the same, then there is no reduction in explosives instantaneously detonated.

Using the shot in Figure 10.7, let us assume that the borehole diameter is 3 in (73.5 mm), the burden and spacing is 6 ft $\times$ 7 ft (1.82 m $\times$ 2.13 m), and the loading density of the explosive is 0.82 g/cm^3. If the borehole diameter is reduced to 2 in (50.8 mm) the other dimensions will change. To maintain the original powder factor we must first take the loading density of the explosive and compare the remaining borehole sizes. The loading density of 0.82 g/cm^3 will yield 2.51 lb of explosives per linear foot (3.73 kg/m) in the 3-in (73.5-mm) borehole and 1.12 lb per linear foot (1.67 kg/m) in the 2-in (50.8-mm) borehole. Making a ratio of these two borehole yields we obtain

$$\frac{2.51}{1.12} = 2.25$$

Next we must calculate the volume of rock yield from the original burden and spacing and from that determine the new burden and spacing.

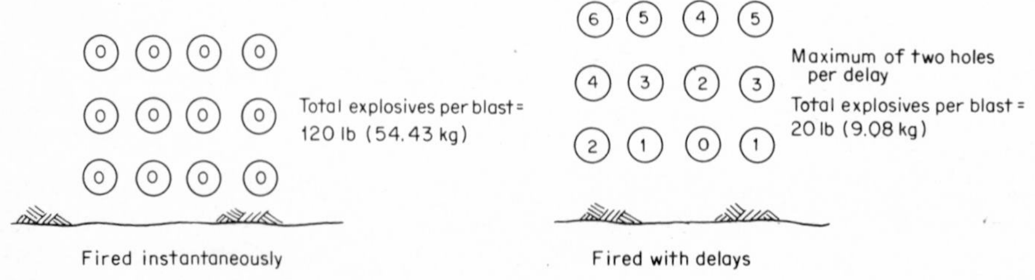

Fig. 10.7 Firing simultaneously and with delays.

TABLE 10.4 Cubic Yards of Rock per Foot of Borehole

	Spacing, ft							
Burden, ft	3	4	5	6	7	8	9	10
3	0.33	0.44	0.56	0.67	0.78	0.89	1.00	1.11
4	0.44	0.59	0.74	0.89	1.04	1.19	1.33	1.48
5	0.56	0.74	0.93	1.11	1.30	1.48	1.67	1.85
6	0.67	0.89	1.11	1.35	1.56	1.78	2.00	2.22
7	0.78	1.04	1.30	1.56	1.81	2.07	2.33	2.59
8	0.89	1.19	1.48	1.78	2.07	2.37	2.67	2.96
9	1.00	1.33	1.67	2.00	2.33	2.67	3.00	3,33
10	1.11	1.48	1.85	2.22	2.59	2.96	3.33	3.70

The volume of rock per foot of borehole for a burden and spacing of 6 ft × 7 ft is 1.56 yd³/ft. To maintain the powder factor the ratio of loading densities must be applied to the rock yield. Therefore

$$\frac{1.56 \text{ yd}^3/\text{ft}}{2.25} = 0.69 \text{ yd}^3/\text{ft rock yield required}$$

This required rock yield of 0.69 yd³/ft is then compared to Table 10.4 to determine the burden and spacing required to produce this volume. We find that the burden and spacing that is closest in yield is 3 ft × 6 ft (0.91 m × 1.82 m), which yields 0.69 yd³/ft.

Converting this burden and spacing into the number of additional holes gives a new total of 35 boreholes. If we had assumed in the original example a delay shot of 120 lb (54 kg), then instead of 10 lb (4.53 kg) per hole the shot would contain 3.43 lb (1.55 kg) per hole. Therefore, if one were to blast one hole at a time, the explosives blasted instantaneously would be reduced by 34%.

Benching: Benching is another method in which vibration due to blasting can be reduced. Benching is the reduction or decrease in the depth of cut whereby to obtain the required depth of cut smaller cuts are used totaling to the final depth. For example, if there is a highway cut 60 ft (18.3 m) in total depth, this may be blasted by benching cuts of 30 ft (9 m) or less (see Figure 10.8). In this method, the top 30 ft would be drilled, blasted, and excavated and then the process of drilling and blasting would be repeated on the bench, or level. By cutting the individual depth of cut in half, the amount of instantaneous explosives is cut in half. (In this example it may even become necessary to reduce the benches to obtain the final grade.)

It is sometimes prudent to reduce the number of holes fired on the first shot. The first shot is often the shot that has to "open" the face for the rest of the blast; it has to move rock that is more restricted to make a free face for the remaining shots to break toward. Therefore, if the first shot is reduced in

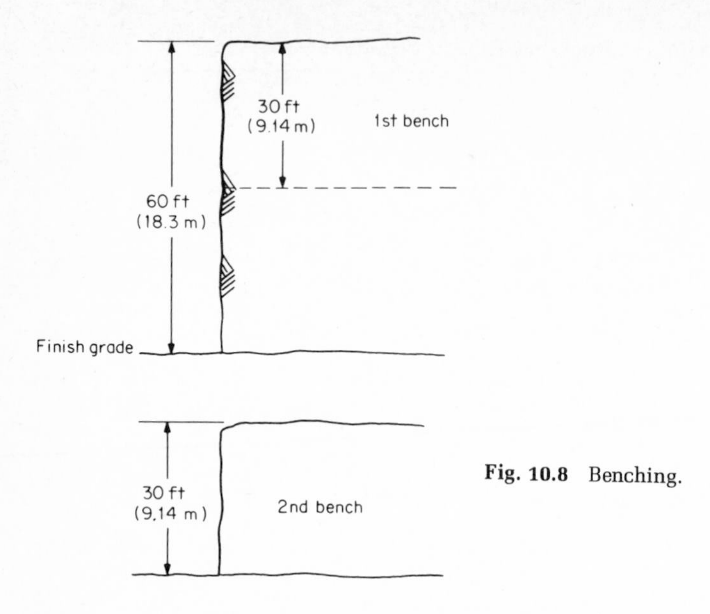

Fig. 10.8 Benching.

pounds per delay it will help reduce vibrations. Moving rock for a free face generally creates more vibration because of wasted energy, since, when there is not a good free face, one must be created. Creating this free face often requires a greater powder factor; also, the rock offers additional resistance, because it has no direction to move freely. Therefore, greater care has to be exercised in designing the initial shot.

Decking: Decking can also be used to reduce the amount of explosives per blast. Decking is best used when it is impractical to bench; however, the length of the column of explosives must be reduced. The column of explosives may be segmented into several decks of different delay caps, thereby reducing the amount of explosives detonated at the same time.

Line Drilling: Another method of reducing vibration, in a particular direction, is to line-drill between the blast area and the structure that is being protected from vibration. (Line drilling was discussed in detail in Chapter 8.)

To conclude our discussion of blasting vibrations, it must be realized that the methods discussed here are only a few of the basic methods for decreasing vibrations from blasting. These methods can be used singly or in any combination, depending on the seriousness of the vibration problem and a cost-risk analysis. The main point to keep in mind is that most blasting vibration problems can be solved when good blast design principles are followed.

Flyrock

Flyrock is the undesirable throw or movement of rock or debris from the blast area. Good blast design is the primary method of avoiding flyrock; however, good blast design often is not enough, and the blaster must resort to methods of containing the flyrock. There are two primary methods of covering or containing the flyrock: backfilling and blasting mats.

Backfilling Backfilling is covering a blast with soil, preferably sand, to control or prevent undesirable flyrock. The rule of thumb for backfilling is that the amount of backfill is equal in depth to the amount of stemming; however, backfill must be a minimum of 3 ft (0.9 m). For example, a shot in which one is holding 3 ft of stemming would require at least 3 ft of backfill. (See Figure 10.9.)

When backfilling, a blaster must exercise extreme caution not to break the cap leg wires. The best way to prevent this is to hand-shovel fine material, burying the wires to protect them from being cut or broken by coarse material during backfilling. While one worker places the backfill with a machine, another person should monitor the circuit with a galvanometer. In this way the blaster can pinpoint when a wire is broken. During the backfilling operation the blaster should direct the operator where and how fast to place the backfill. This also enables the blaster to know the approximate location of a broken wire. While the blaster is watching the backfill being placed, the person monitoring the galvanometer can indicate to the blaster when a break occurs. At this time the blaster can stop the backfilling operation and find the broken wire with relative speed.

Backfilling is generally most successful in a trench, because a trench generally requires a greater powder factor, and being narrow it lends itself to being backfilled rapidly. In backfilling something other than a narrow trench, often the machine being used has to maneuver on the shot. This can be done without much problem provided care is taken to obtain the maximum depth of a minimum of 3 ft (0.9 m) before the machine walks on that area of the shot.

Backfilling is advantageous in that it requires less equipment than blasting mats, it can produce better breakage, and the whole blast can be shot at once. An example of the reduced equipment requirement is doing a trench for a utility contractor; often the contractor has a backhoe or loader there for laying pipe. In this case the blaster needs only a powder truck, whereas if blasting mats were used, the blaster would also need a crane to handle the

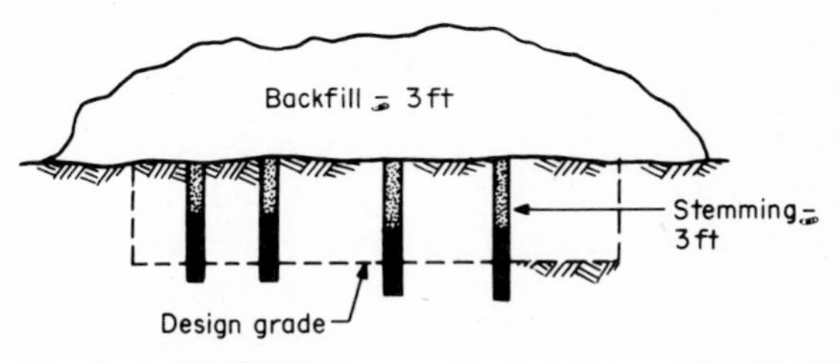

Fig. 10.9 Backfilling.

Fig. 10.10 Blasting mat prohibiting flyrock from traveling.

mats and a truck to transport the mats. (Producing better breakage and shooting the entire shot at once can also be the main disadvantage.)

Using backfill for blast coverage generally requires more explosives, because the backfill contains the movement of the rock. The best results can be achieved by holding approximately one-third the original stemming. For example, a shot that was originally designed with 3 ft (0.9 m) of stemming should have the stemming reduced to 1 ft (0.3 m). The increased powder factor not only improves breakage but also increases the problems of vibration, which is also increased by the restriction to blasting the whole shot at once because the larger number of holes would require blasting more holes on one delay. This affects the amount of powder per delay, therefore increasing vibration.

Blasting Mats Blasting mats are netting or matting constructed of either cable or rubber tires and designed to contain the rock or prevent it from flying when blasted (see Figure 10.10). The common steel-cable blasting mats measure 10 ft × 12 ft (3m × 3.65 m) and weigh approximately 3000 lb (1365 kg). When the blast holes are loaded and wired into the lead wire, one or several blasting mats are placed on the blast, covering all the holes to be blasted. The cost of blasting with mats is greater and the speed is less than with traditional open shooting. The main rule for blasting mats is never to blast more holes than can be safely covered with the mats.

Although blasting mats are heavy, the movement of the rock will often lift the mats enough to permit flyrock. In some cases the blast can contain enough force to throw the mats and cause damage.

As with backfilling, great care should be taken not to break any leg wires in placing the mats. In addition to the precautions of monitoring the galvanometer and covering the leg wires, it is important to place the mat on the area that it is to cover rather than dragging it on the shot. When using wire-rope mat one must make sure that the bare wire connections are insulated from the mat. If the steel comes in contact with the wire it will cause a short, preventing the shot from detonating or, worse yet, causing partial detonation.

Wire-rope and rubber-tire mats function similarly for blast control. Rubber mats offer less chance of cutting or shorting out the wires and tend to be better suited for narrow trenches; however, because of the much greater weight of rubber mats, wire-rope mats are often more desirable. Also, wire-rope mats are more flexible and so can match the contour of the shot more readily.

When using mats to cover very shallow holes (less than 2.5 ft, or 0.76 m) it is advisable to cover the borehole with plywood or an old tire to prevent damage to the mat. Shallow holes have a greater tendency to rifle and blow holes in the mat.

Backup Holes: When mats are used it will generally require more than one shot to blast the whole formation. The surrounding conditions may require the blaster, in some cases, to limit the blasts to one or two holes per blast. Because of the costs of drilling, generally for small jobs (i.e., house foundations, swimming pools) the drilling is completed before the blasting has begun. In blasting where all the drilled holes cannot be shot at once, there is a danger of causing rock movement, destroying the loadability of an adjacent hole. This necessitates the loading of "backup" holes. To backup a shot is to load the holes adjacent to the holes to be blasted. If the blaster uses electric caps to load the backup holes, there is a good chance that the movement of the rock will cut the cap leg wires. If this happened, the blaster would be faced with a primed charge at the bottom of the hole with the means of detonating the charge lost. The best method of loading a backup hole is to use detonating cord. Detonating cord is strong enough to resist most rock movement, but if the cord is severed, the charge can be excavated with relative safety.

It is not recommended to load a backup hole with the same amount of explosives as other holes. It is very likely that when the shot is fired the rock will break to the backup hole. If the backup holes of a shot with holes 6 ft (1.8 m) deep that require 3 ft (0.9 m) of stemming also have 3 ft (0.9 m) of stemming, this may leave the backup hole with less than the desired 3 ft (0.9 m) to the nearest free face. (See Figure 10.11.)

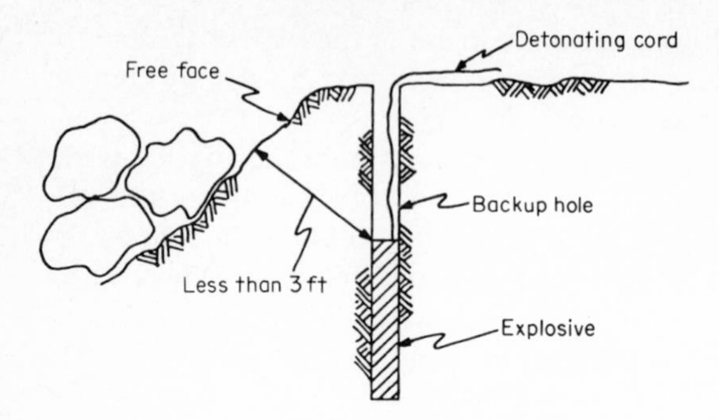

Fig. 10.11 When loading backup holes, do not use as much powder as in the other holes.

The best method for loading a backup hole is to reduce the amount of explosive by one-third to one-half and add approximately 1 ft (0.3 m) of stemming. After the blast, the blaster can determine if the hole requires any additional explosives. If the explosives are required, the blaster may determine how much explosive to add to obtain the desired distance to the free face. Since detonating cord is used in the backup hole, the blaster has only to add the explosives, because the cap will be tied to the detonating cord at the surface. When a shot back-breaks, the surface rock will occasionally break in

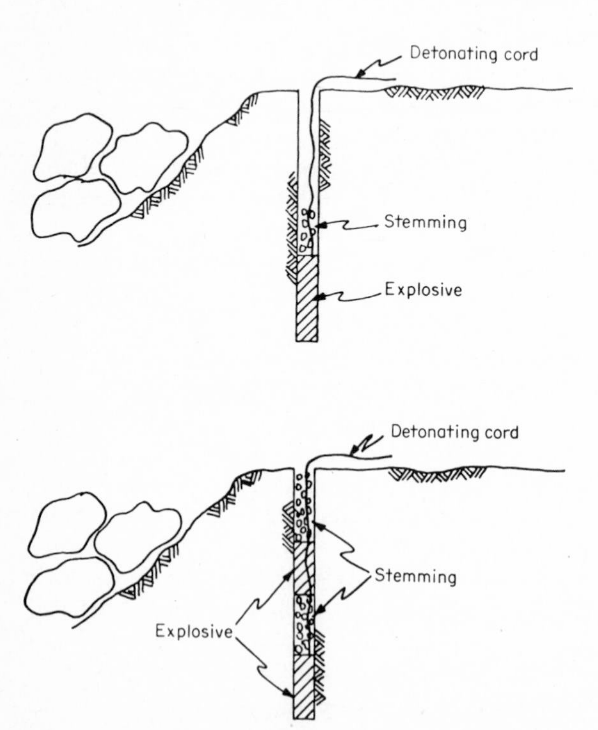

Fig. 10.12 Blaster may add explosives to maintain distance to the nearest free face.

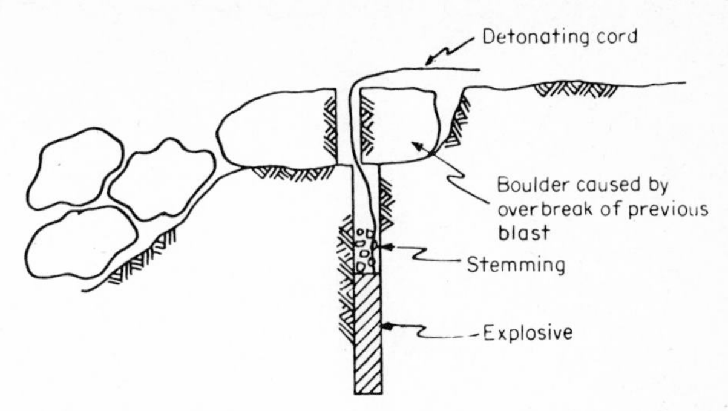

Fig. 10.13 Blaster may add explosives to boulder caused by back break.

the form of a boulder. Using the method of backup loading will permit the blaster to place a small amount of explosives into the boulder to break it with the rest of the hole. (See Figures 10.12 and 10.13.)

After the backup hole is completely loaded, the blaster must either attach the detonating cord to a detonating cord trunk line or attach a blasting cap to the detonating cord. When a cap is attached to a single hole with detonating cord it should be attached as close to the top of the borehole as possible. The explosive end of the cap (the opposite end from the leg wires) should be the closest to the hole. The cap must be securely attached to the detonating cord. Generally electrician's tape wrapped around the cap and the cord is one of the best methods of securing the cap and the detonating cord together (see Figure 10.14). After the cap is attached to the detonating cord, the excess cord should be cut between 3 and 6 in (76 and 152 mm) from the cap. After the cap is tied into the rest of the shot, the cap and detonating cord should be covered with sand or other soft earth. This protects the cap from damage or

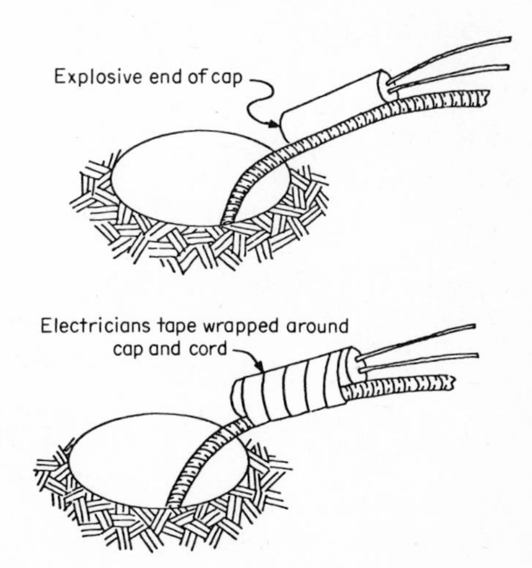

Fig. 10.14 Proper method for attaching cap to detonating cord.

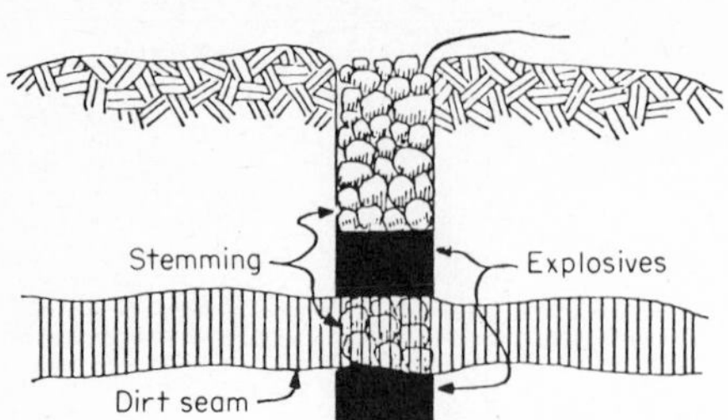
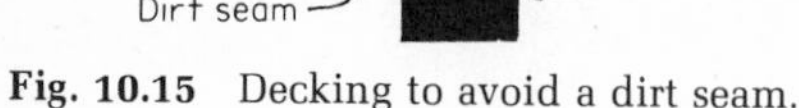

Fig. 10.15 Decking to avoid a dirt seam.

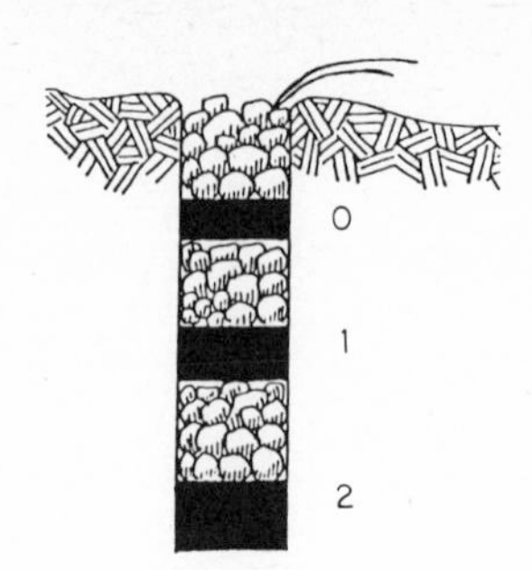

Fig. 10.16 Decking using different delays.

accidental detonation from the mat and helps to cut the noise created by the detonation of the detonating cord. It is most imperative that all backup holes be completely covered with mats. If there is a mud seam within the rock the blast may cause one of the backup holes to detonate with the shot. Therefore, as a safety precaution the backup holes must be matted.

Whether backfill or mats are used for control, there is no replacement for a good blast design. Only the proper use of these coverings enables the blaster to successfully attempt any rock blasting project.

DECKING TECHNIQUE

Decking is a technique of loading whereby the column of explosives is interrupted by a layer of stemming. This technique enables the blaster to avoid seams in the formation and to bring the explosives closer to the top of the hole without increasing the total amount of explosives, and it permits the distribution of the explosives to those areas of the formation where the concentration of explosives is most needed. The deck may be used to bypass seams in the rock formation that may permit the escape of gases. (See Figure 10.15.)

In decking, each deck must have contact with a detonator. With detonating cord, the cord is the complete length of the hole and so this is not a problem. However, if caps are being used, there must be a cap with each deck. If caps are used, vibration can again be reduced by using a different

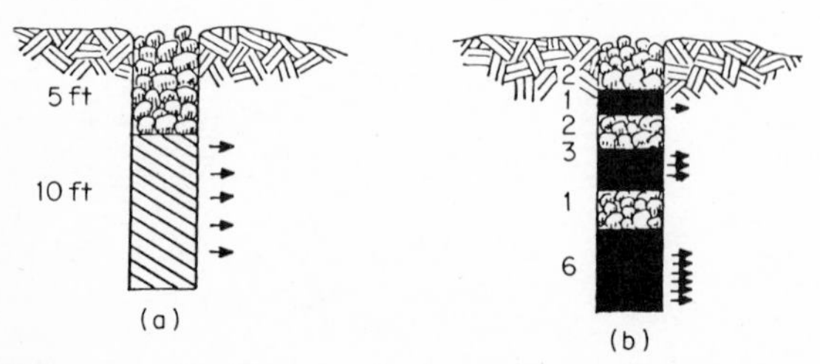

Fig. 10.17 Decking to reduce the total explosives used.

delay for each deck. However, the delays must be so arranged that the top deck is the earliest delay. (See Figure 10.16.)

Often, when vibration is of concern, the use of decks enables the blaster to distribute the explosive charges throughout the borehole (covering more gross area) while decreasing the amount of total explosive. Figure 10.17a shows a hole loaded without a deck. The explosives are distributed in the bottom two-thirds of the borehole. In Figure 10.17b, the same amount of explosives is distributed over the bottom 13 ft (4 m). Of course this decking is only needed when there is a concern with vibration. With the use of the deck, the top of the hole will have better fragmentation, but it must be realized what is gained in energy near the top of the borehole is taken from the bottom.

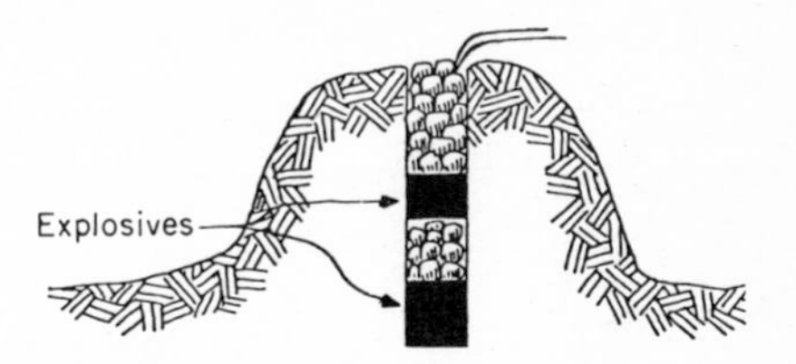

Fig. 10.18 Decking to suit terrain.

A deck can permit the distribution of explosives where the concentration is most needed. In Figure 10.18 the stemming must be increased to keep the distance from the nearest free face. However, using that amount of stemming will obviously leave a boulder on the top of the blast. If a deck is used, the stemming may still be increased, yet the blaster can prevent a boulder by putting a charge in the center of mass of the potential boulder.

To conclude, it has been stressed that the most important factor in reducing vibration and controlling flyrock is good blast design. That is why it is imperative that, when blasting in areas where vibrations and flyrock can create problems, those responsible for blasting operations make a conscious effort toward a good blast design.

ELEVEN

SPECIAL BLASTING PROBLEMS

TRENCHES

Because of their geometric dimensions and their usual close proximity to structures, trenches tend to be one of the most challenging kinds of blasting projects. Trenches are generally narrow, requiring the movement of rock along the axis of the trench; i.e., the shape of a trench requires the movement of rock several burdens in length at the axis whereas the width is just one or two spacings. This necessitates increasing the powder factor. For most rock projects, the powder factor in pounds per cubic yard (lb/yd^3) ranges between 0.75 and 1.25. However, with trench blasting, the powder factor often averages in the range of 2–3 lb/yd^3. Generally, the spacing in a trench for a 1¾-in (44-mm) hole will not exceed 3 ft (0.9 m), whereas the burden may reach 4 ft (1.2 m). On ordinary rock cuts the burdens do not exceed the spacing; however, in trench work this may occasionally happen. (See Figure 11.1.)

Trenches with a depth of cut in excess of 8 ft (2.4 m) are generally drilled with a crawler drill with hole diameters of 2 in (50 mm) or more; however, trenches with cuts less than 8 ft (2.4 m) are often drilled by hand with 1½- to 1¾-in (38- to 44-mm) holes. The small-diameter hole is required when smaller drills are used, but it also permits a close spacing, which is favorable because many trenches do not exceed 2 ft (0.6 m) in width.

Trench Drilling Patterns

There are four basic drilling patterns that may be used for trenches. In Figure 11.2 the double row of holes (*a*) is probably the most common drill pattern. The burden and spacings are generally equal and the delays are usually the same for holes opposite each other. Figure 11.2*b* consists of one single row of drilling in the center of the trench. This drill pattern is limited to medium and soft rock and is not recommended for hard rock types.

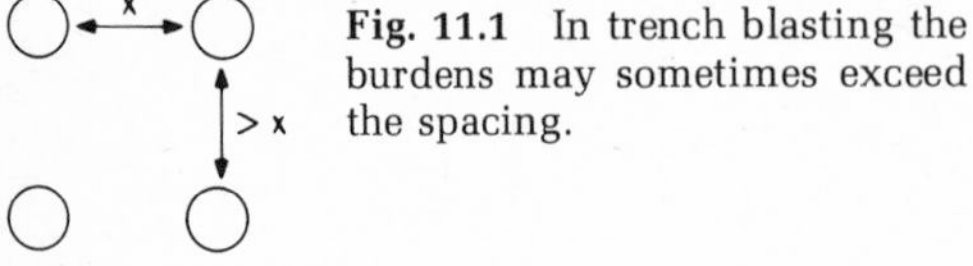

Fig. 11.1 In trench blasting the burdens may sometimes exceed the spacing.

The drill pattern in Figure 11.2*c* can be used for trenches that are equal to twice the spacing in width. A trench that is 5 ft (1.5 m) wide and is being drilled with a bit 1½ in (38 mm) in diameter would be an excellent application of this drill pattern. The spacing could be at 3 ft (0.9 m), in which case the trench would be drilled 6 ft (1.8 m) wide. In designing the delay system for this drill pattern, it is recommended that the center hole fire one delay period earlier than the hole to either side. In addition to this delay design, it is generally a good practice to load the center hole with more explosives than holes to the side. This combination of delay design and loading techniques will cause the rock to move out of the center first and will permit the rock from the side holes to move towards the center of the trench. This results in a muck pile that is easy to excavate. (See Figure 11.3.)

The drill pattern in Figure 11.2*d* is generally used for narrow trenches. This pattern enables the burdens to be extended, lowering the powder factor, yet the center hole prevents the burdens from becoming large. Often the center hole, called a "reliever," is shallower than the other holes. The shallow center hole is used to help the movement of the surface rock to give way for the deep rock. For example, if the holes on the sides of the trench are 4 ft (1.2 m) deep, then the addition of a 2-ft (0.6-m) center hole will permit the spreading of the burdens by the use of the additional explosives for the top

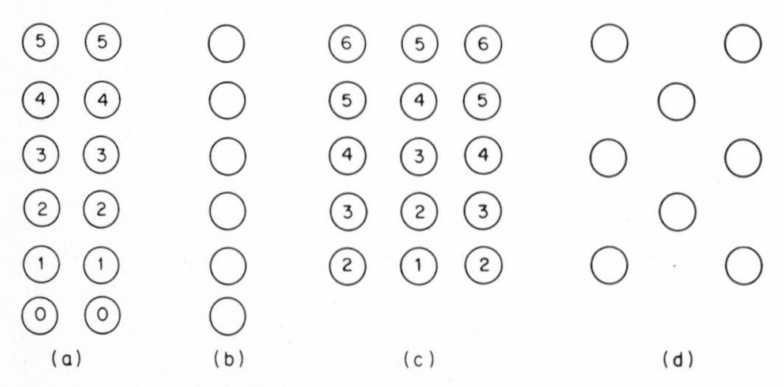

Fig. 11.2 Trench drill patterns.

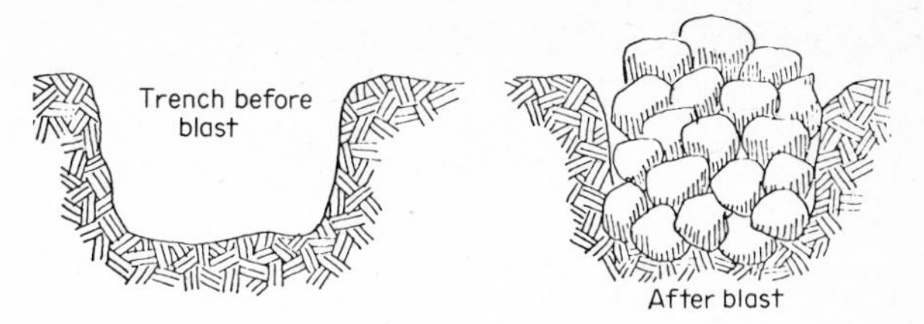

Fig. 11.3 Trench muck pile.

rock. This method can be particularly effective in avoiding seams used in conjunction with, or in lieu of, decking.

Delay Design for Trenching

Generally the delay design in trenches is such that the blast is aimed, or delayed, to move in one particular direction. This usually is away from a source of liability, such as a building or pipe; it may also be in the direction of a natural free face. However, in some situations the delay may be from the center. For working between sheathing beams with a clam-type excavator a delay design that has the middle of the shot detonating first has been very effective in aiding the excavation. Using a clam between sheathing beams requires good fragmentation, but the placement of the muck pile must be between the cross members to enable the clam to dig the muck. Also, the drill pattern in this type of situation may resemble a round using a pyramid cut at the middle point between the cross members (see Figure 11.4).

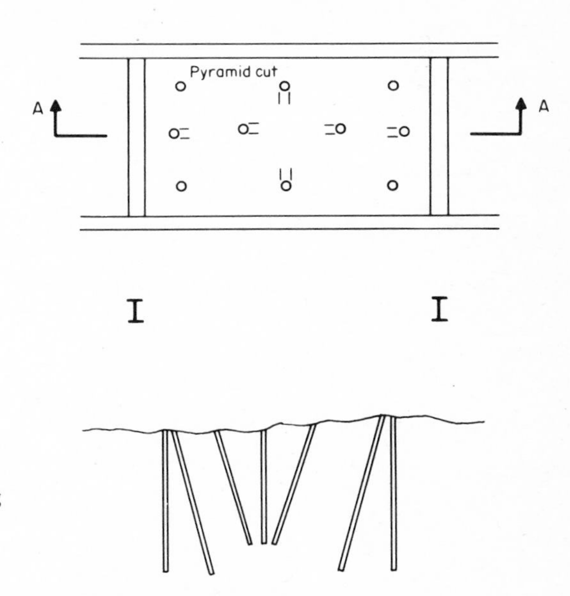

Fig. 11.4 Pattern for blasting deep trenches.

TABLE 11.1 Trench Stemming

Trench depth, ft*	Drilling depth, ft	Spacing, ft	Stemming, ft
1	2.5–3	2	2
2	3.5	2.5	2.5
3	4.5	3	3
4	5.5	3	3
5	6.5	3	3
6	7.5	3.5	3.5

* For metric, multiply by 0.3048.

The delay design for this technique may cause flyrock; if the trench is quite deep (which can be assumed with sheathing) then flyrock may be easily controllable with blasting mats.

Trench Burdens and Subdrilling

Table 11.1 consists of recommended depths and burdens for trenches. Keep in mind that a good rule of thumb for subdrilling a trench is one-half the burden distance. The stemming usually equals the spacing.

Ditching

Ditching is an effective way of excavating a trench in moist or wet loam-type soils, provided the soil is not cohesive. Trenches ranging from 2 to 12 ft (0.6 to 3.65 m) deep and 4 to 40 ft (1.2 to 12.2 m) wide can be quickly excavated with this method. Ditching can reduce excavating costs, it is fast and simple, and it permits the excavation of trenches in areas often unworkable by mechanical equipment.

When ditching, it is advisable to make small test shots to obtain an understanding of the explosives requirements.

The Single-Row Method One method of ditching requires a single line of holes, made with a bar or rod, uniformly spaced at even elevations. In uneven terrain a blaster should vary the depths of the holes but place the charges at the same elevations. (See Figure 11.5.) Table 11.2 gives specifications for blasting with the single-row method.

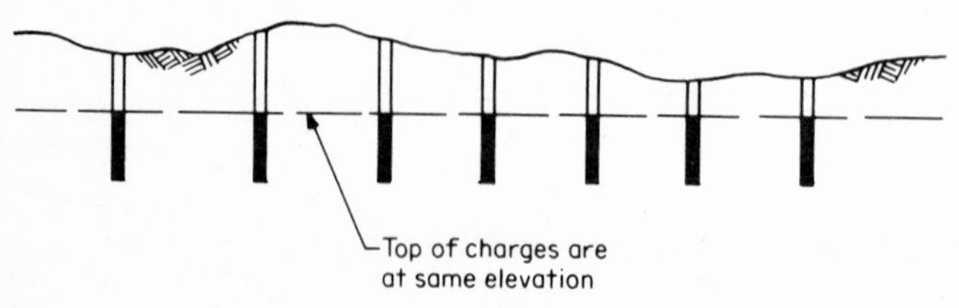

Fig. 11.5 Explosives must be placed at same elevation in ditching.

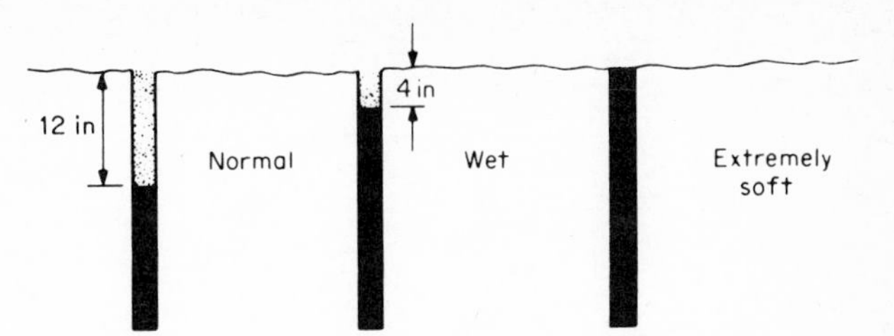

Fig. 11.6 Stemming requirements for ditching.

The minimum stemming for these holes is 1 ft (0.3 m), with only 4 in (101 mm) of stemming in extremely wet ground. If the material is extremely soft, there is no stemming used at all. In this case the explosives are brought even with the surface. (See Figure 11.6.)

In a single-row trench it is recommended that the end of the trench have a charge on each side of the trench instead of a single charge in the center. This is because the last hole tends to give a round effect that placing the charges on each side avoids.

This single-row method is best suited for small drainage ditches.

The cross-sectional method of ditching, shown in Figure 11.7, is a single-row setup with a line of holes at 90° to the single row at every other hole. Table 11.3 shows specifications for the cross-sectional loading method.

The Posthole Method The posthole method, shown in Figure 11.8, is for ditches that are deeper than 5 ft (1.52 m). A posthole auger is used to drill the hole. Instead of a column of explosives, a bulk of explosives is placed at the bottom of the hole. If wide-diameter charges are not available one can tie several small (1¼- × 8-in) charges together.

The charge is placed at two-thirds the required depth of the trench. This method will yield a trench with a bottom width equal to the depth and a top

TABLE 11.2 Specifications for Ditches with a Single Line of Charges

Cartridges per hole (1¼ in × 8 in)	Depth to top of charge, in	Distance between holes, in	Probable depth of ditch, ft	Probable top width of ditch, ft	CIL ditching dynamite, lb/100 ft
½	6–8	12	1½	4	25
1	6–12	15	2½	6	40
2	6–12	18	3	8	65
3	6–12	21	4	10	85
4	6–12	24	5	13	100
5	6–12	24	6	16	125

SOURCE: "Blasters' Handbook," Explosives Div., Canadian Industries Limited, Montreal, 1968, p. 346.

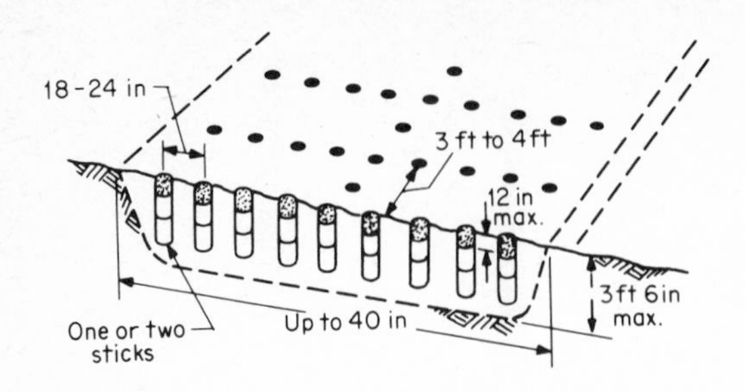

Fig. 11.7 Cross-sectional method of loading to clean and widen ditches. (*U.S. Army.*)

width equal to 3 times the depth. A 3-lb (1.36 kg) charge will generally propagate 3 ft (0.9 m) (depending on soil conditions); however, a test shot is recommended. In general for a trench greater than 6 ft (1.82 m) deep, large charges will yield more dirt moved per pound of explosives than the single-column method.

Table 11.4 gives specifications for the posthole method.

The Relief Method The relief method is used where heavy sod or roots form a mat over the trench area that will resist the ditching. This necessitates cutting the mat at the side of the trench. This can be done with a spade or an ax or by plowing a furrow with mechanical means. The charges are placed at the edge of the trench in a method similar to the single row. However, these charges are less than half the weight of the main charges. (See Figure 11.9.)

Blasting Technique In ditching, the holes must be vertical and the elevation of the charges must remain constant for a uniform cross section. Ditching holes, unless underwater, must always be stemmed. It is important to

TABLE 11.3 Specifications for Ditches with Cross-Sectional Loading Method

	Cartridges per hole (1¼ in × 8 in)	1	2	3	4	5
Holes	Hole spacing, in	15	18	21	24	24
per	Row spacing, in	30	36	42	48	48
row	Depth of ditch, ft	2½–3	3–3½	4–4½	5–5½	6–6½
3	Width, ft	11	11	13	17	20
	Dynamite, lb/100 ft	80	133	172	200	250
5	Width, ft	—	14	17	21	24
	Dynamite, lb/100 ft	—	200	257	300	315
7	Width, ft	—	—	20	25	28
	Dynamite, lb/100 ft	—	—	343	400	500
9	Width, ft	—	—	—	29	32
	Dynamite, lb/100 ft	—	—	—	500	625
11	Width, ft	—	—	—	—	36
	Dynamite, lb/100 ft	—	—	—	—	750

SOURCE: "Blasters' Handbook," Explosives Div., Canadian Industries Limited, Montreal, 1968, p. 347.

TABLE 11.4 Specifications for Ditches with Posthole Loading Method

Charge per hole, lb	Depth of charge, ft	Distance between holes, ft	Probable depth of ditch, ft	Probable bottom width, ft	Probable top width, ft	CIL ditching dynamite, lb/100 ft
3	2½	3	4	4	12	100
5	3	3½	5	5	15	140
10	4	4	6	6	18	250
15	4½	4½	7	7	21	335
25	5½	5	8	8	25	500
50	8	6	12	12	36	835

SOURCE: "Blasters' Handbook," Explosives Div., Canadian Industries Limited, Montreal, 1968, p. 349.

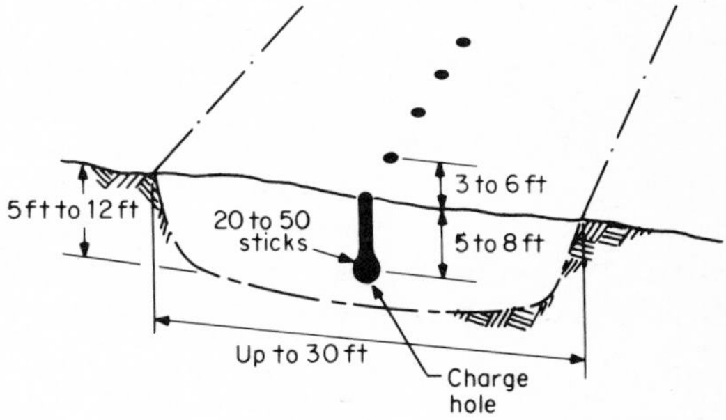

Fig. 11.8 Posthole method of loading for shallow ditches in mud. (*U.S. Army.*)

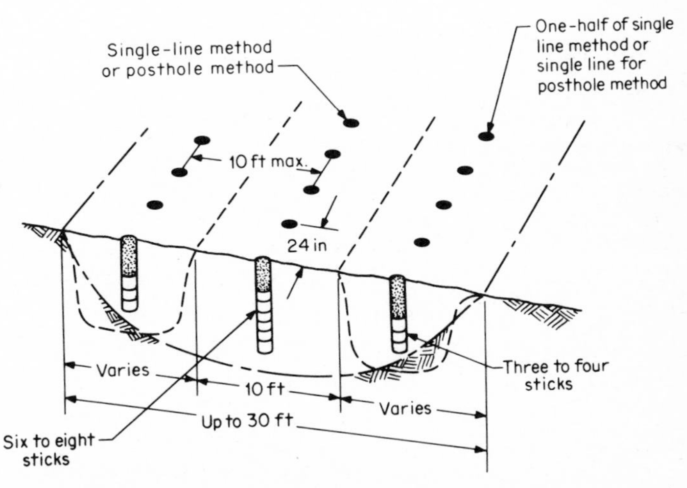

Fig. 11.9 Relief method of loading for shallow ditches. (*U.S. Army.*)

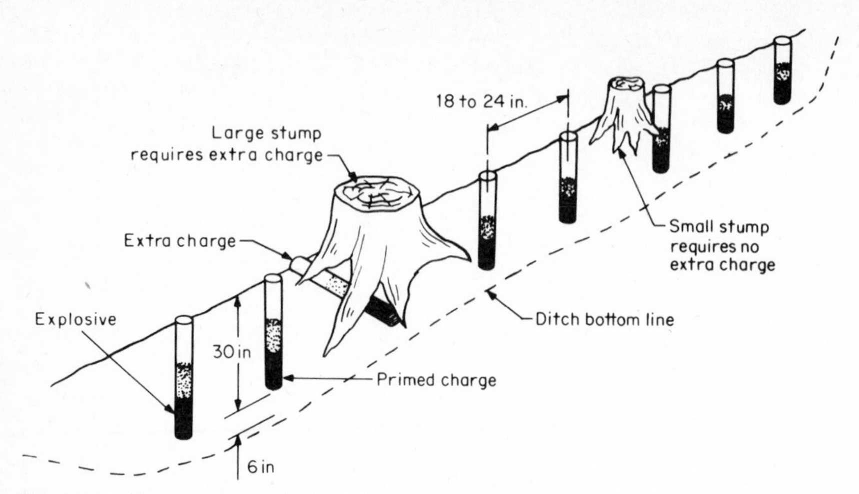

Fig. 11.10 Propagation method of detonation. (*U.S. Army.*)

know the kind of soil and the depth of the various strata. Generally the heavier the soil, the deeper the charge; and the greater the water, the shallower the charge. The water depth is treated as soil depth.

Always remove branches that overhang the trench, because they will cause the blasted material to fall back into the trench. If stumps, boulders, or logs are in the intended trench line, remove them at the same time the trench is blasted.

The ditch blast may be fired by propagation, caps, or detonating cord. Propagation is generally successful in wet soils with sensitive explosives. The moisture content of the soil must be high enough so that when you squeeze a soil sample in your hand it can be molded into a shape that will stay.

Propagation is the cheapest and simplest method of detonation. One charge must be primed with a cap or detonating cord. If there is more than one row a blaster should prime one charge in each row (generally the end charge). The propagation distance varies with conditions and the amount of explosives. In warm, moist soil a quarter pound of explosives will propagate about 1 ft, a half pound will propagate approximately 18 in, and 50 lb will propagate 10 ft. (See Figure 11.10.)

If the conditions are not favorable for propagating then each hole must be primed with detonating cord or caps of the delay.

COYOTE BLASTING

In some cases, the most economical means of blasting heavy side hill cuts in very rough terrain or quarries for riprap or jetty stone is the coyote method. A coyote blast is a series of small tunnels driven into the toe of the quarry face and loaded with explosives. The simplest coyote consists of a horizontal tunnel (adit) driven into the center of the toe perpendicular to the face; this tunnel is called the main stem. The main stem then has a wing or crosscut adits, called wings, driven from it at 90° to its axis. The powder distribution for the method is very poor, but it works well in closely jointed formations or for producing riprap.

For best results, the quarry face should be 75 to 100 ft (23 to 30.5 m) high; if higher than 100 ft (30.5 m), a second set of wings is required. The length of the main stem is 0.6 to 0.75 times the quarry face height. (See Figure 11.11.)

Once the wings are loaded, the main stem is backfilled with crushed stone screenings (preferably) as stemming. Because of the hazards, blasting agents are used instead of dynamites.

Powder and Shear Factors

The powder factor and shear factor are the determinants of the amount of explosive to be used. Shear factor is the weight of the explosive in pounds divided by the area of the base of the shot in square feet. Single-wing shots require a shear factor of 2.0 to 2.5 lb/ft^2 (9.76 to 12.21 kg/m^2), whereas a

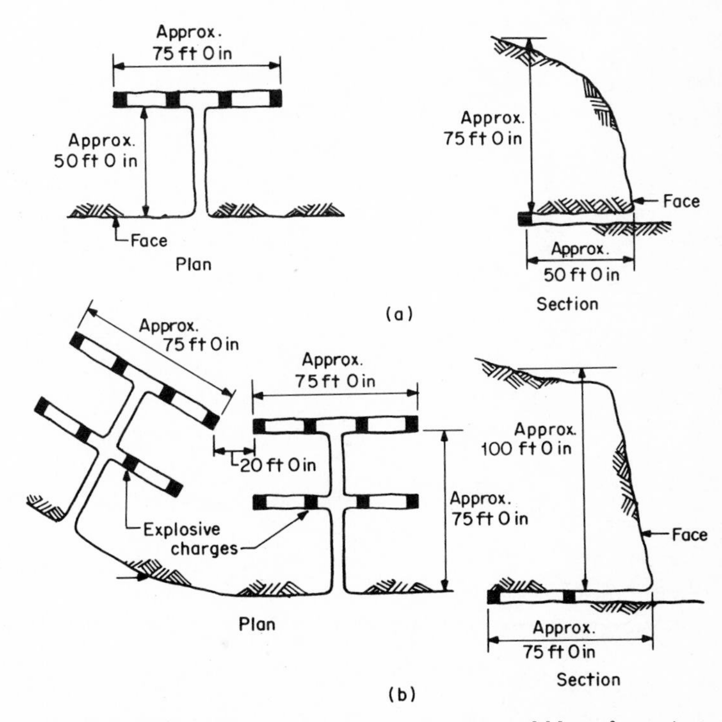

Fig. 11.11 Representative layouts of coyote or tunnel blasts for various face heights: (*a*) single tunnel and (*b*) two double tunnels. (*Canadian Industries Limited.*)

multiple-wing shot with a bank height that does not increase too rapidly requires a shear factor at the first wing of 2.5 to 2.75 lb/ft² (12.21 to 13.4 kg/m²) and 3.5 to 4.0 lb/ft² (17.1 to 19.5 kg/m²) at the back wing.

The normal powder factor for the single T is 0.75 to 1.25 lb/yd³ (0.45 to 0.74 kg/m³), whereas for multiple T's, or multiple wings, the powder factor for the front T is 1.0 to 1.25 lb/yd³ (0.59 to 0.74 kg/m³) and for the back T 1.25 to 1.50 lb/yd³ (0.74 to 0.89 kg/m³) for an overall average of 1.2 to 1.4 lb/yd³ (0.71 to 0.83 kg/m³) for the total shot.

Generally the muck pile will be less than 1.5 times the face height; however, for planning the shot the multiple 2.0 should be used. (See Figure 11.12.)

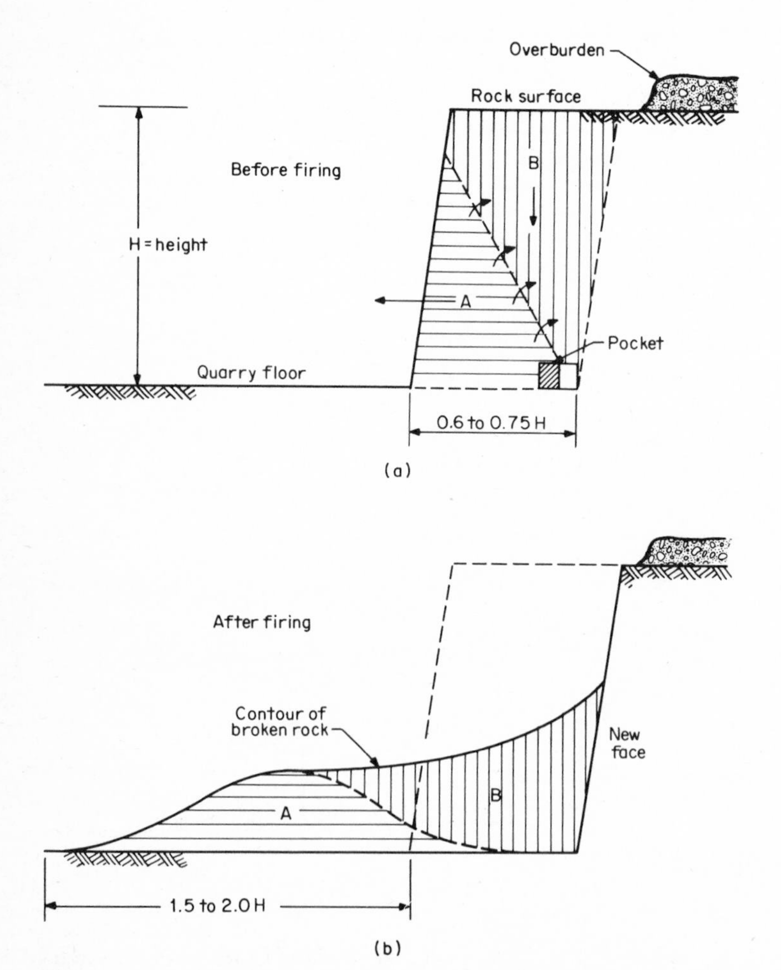

Fig. 11.12 Mechanics of coyote tunnel blast: (*a*) before firing and (*b*) after firing. (*Canadian Industries Limited.*)

BOULDERS

Boulders can be formed from nature or caused by insufficient breakage during blasting; the latter are called "secondary." Boulders can be broken for easier handling by blockholing, mudcapping, or snakeholing.

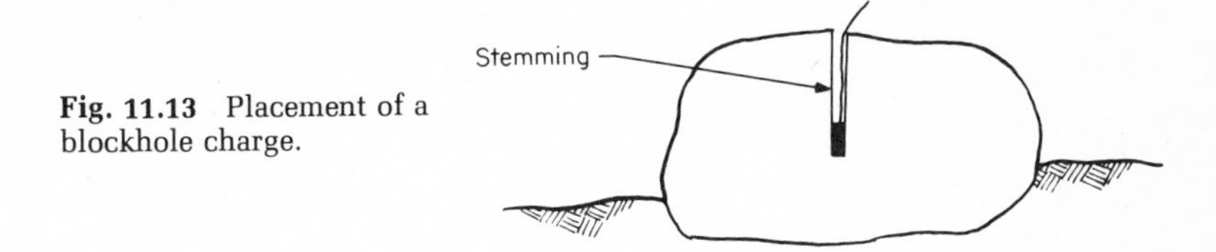

Fig. 11.13 Placement of a blockhole charge.

Blockholing

"Blockholing" refers to the method of breaking boulders that requires a hole drilled in the center of mass to a length from one-half to two-thirds the boulder's depth. Larger boulders may require more than one hole. Usually one can figure one hole per 8 to 12 ft^2 (0.7 to 1.1 m^2) of horizontal cross section at the boulder's widest spot; i.e., a boulder that is 8 ft (2.4 m) in diameter at the widest dimension would require on the average $\pi(4\ ft)^2/(10\ ft^2)$, or five holes.

It is desirable to have one hole in the center of the mass of the boulder about which the other holes are evenly distributed. The powder factor of 0.2 lb/yd^3 (0.12 kg/m^3) will generally give the best results. All holes should be stemmed with peastone or damp sand. Damp, coarse-grained sand, when tamped in the holes, makes the best stemming for boulders. (See Figure 11.13.)

When a boulder is encountered that is partially buried and the actual depth is unknown, the best way to determine how to load it is to drill a hole in what is believed to be the center of mass. This should be continued until the boulder is drilled through. Once this has been done, stem the hole until the net depth of the hole equals one-half to two-thirds of the boulder's depth; then load the boulder accordingly.

Mudcapping

Mudcapping is a contact method of blasting whereby explosives (sticks of dynamite) are embedded in a thin layer of mud placed on the boulder and covered with at least another 4 in (101 mm) of mud. When detonated the explosives will split the boulder. This method requires a powder factor of 3 lb/yd^3 (1.78 kg/m^3). The mud reduces both the amount of explosives required and the noise. It is best to place the charges on a flat surface or slight depression and make sure that the mud is free of rock or any other object that may become a flying projectile. (See Figure 11.14.)

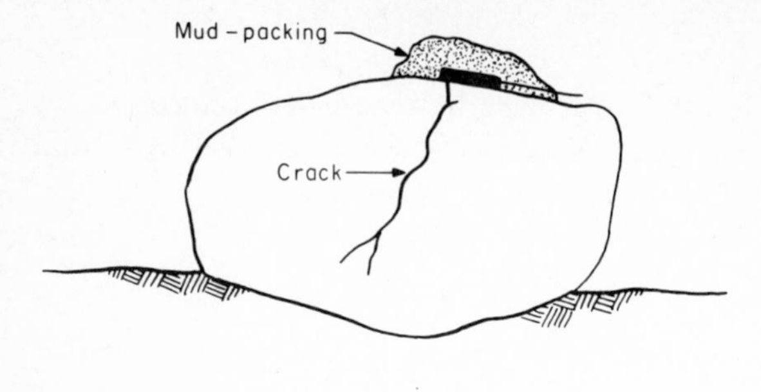

Fig. 11.14 Placement of a mud-capped charge.

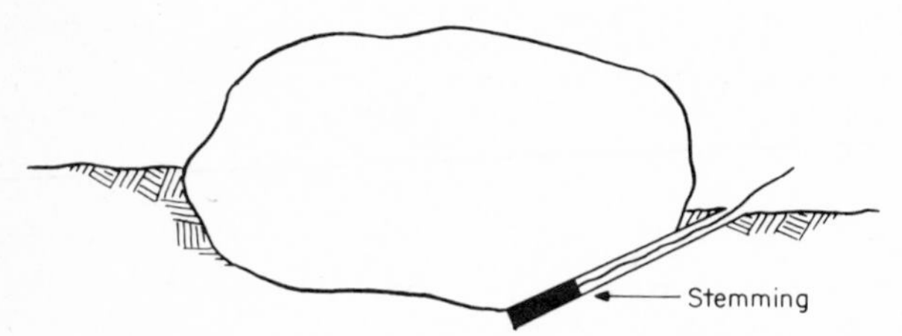

Fig. 11.15 Placement of a snakehole charge.

Mudcapping is restricted in many areas because of the noise and air blast that can be caused.

Snakeholing

The purpose of the snakehole method is to kick the boulder out of the ground and split it at the same time. A snakehole, a hole large enough for the explosive charge, is dug under the boulder (see Figure 11.15). Ensuring that the charge is against the boulder, one should put the primer closest to the surface and bury the hole. The powder factor for snakeholing is 0.5 to 1.0 lb of explosives per ft (0.75 to 1.5 kg/m) of thickness of the boulder opposite the charge.

For larger boulders, a mud cap may be used with a snakehole, provided they are fired simultaneously.

Table 11.5 gives the charge sizes for all three methods of blasting boulders.

TABLE 11.5 Charge Size for Blasting Boulders

	Explosive required, lb		
Boulder diameter, ft	Blockholing	Snakeholing	Mudcapping
3	1/4	3/4	2
4	3/8	2	3 1/2
5	1/2	3	6

SOURCE: *Explosives and Demolitions,* U.S. Department of the Army Field Manual FM 5-25, 1971, p. 3-32.

CONCRETE

Concrete is one of the more difficult materials to blast because of its nonuniform constitution. That is, the strength of a single mass of concrete can vary, and it may have varying strength of rebar within its mass. The theoretical tensile strength of concrete is approximately 10% of its compressive strength; i.e., concrete that has a compressive strength of 6000 lb/in^2 (41,340 kN/m^2) would have strength of 600 lb/in^2 (4134 kN/m^2) in tension. The real difficulty in blasting concrete comes with the addition of reinforcing steel. For example, if there are two footings of equal size, one without reinforcing steel and the other with resteel with a tensile strength of 20,000 lb/in^2 (137,800 kN/m^2), the blasting requirements would differ as follows:

	Without rebar	With rebar
Size	1 yd^3 (3 ft × 3 ft × 3 ft)	1 yd^3
Area at center of mass	1296 in^2 (9 ft^2)	1296 in^2
Area of rebar	None	6 in^2
Tensile strength due to concrete	600 lb/in^2 × 1296 = 777,600 lb/in^2	600 lb/in^2 × 1290 = 774,000 lb/in^2
Strength due to resteel	0	120,000 lb/in^2
Total tensile strength	777,600 lb/in^2	894,000 lb/in^2

In reality the strength of the concrete would fall short of the theoretical level because of cracking, and thus the steel would be a greater contribution to tensile strength than indicated in this example.

The drill pattern and powder amount vary according to the concrete strength. Using our 1-yd^3 (0.76-m^3) block, a weaker or lean concrete would have a powder factor of approximately 0.15 lb/yd^3 (0.09 kg/m^3) and require a hole for 10 ft^2 (0.93 m^2) of horizontal surface. For a strong mix of nonreinforced concrete the powder factor would increase to approximately 0.25 lb/yd^3 (0.15 kg/m^3) and the drill pattern would remain the same. With reinforcement the factors vary according to the quantity of reinforcement. For lightly reinforced concrete the powder factor may reach 0.5–0.75 lb/yd^3 (0.3–0.45 kg/m^3), but for heavily reinforced concrete the powder factor will exceed 1 lb/yd^3 (0.6 kg/m^3) and the drill pattern will reduce to one hole per 6 ft^2 (0.56 m^2). Keep in mind that these powder and drilling factors are just to demonstrate the changes in requirements of various concrete strengths. When you blast concrete, test shots are very important. Concrete is a very unpredictable material to blast, and therefore nothing should be taken for granted.

In addition to strength the powder factor and drill pattern are affected by geometric shape. That is, the drill pattern for a 12-in (0.3-m) concrete wall 5 ft (1.52 m) high will be quite different from the pattern for a massive footing. The footing would very much resemble a drill pattern for a rock cut where the wall requires a single row of holes spaced about 3 ft (0.9 m) apart. However, the spacings in the holes would not vary that greatly. A footing 30 ft (9 m) long, 10 ft (3 m) wide, and 4 ft (1.2 m) deep will require burdens and spacings of approximately 3 ft by 3 ft (0.9 m × 0.9 m); i.e., one hole per 9 ft^2 (0.8 m^2). The wall, on the other hand, would have a single row at 3-ft (0.9-m) spacings, but the relationship of holes to square feet would be one hole per 3 ft^2 (0.27 m^2).

Slabs require shallow holes at closer spacings. That is, a slab 2 ft (0.6 m) thick would require a hole approximately 18 in (0.45-m) deep at 18-in (0.45-m) spacings.

It should be apparent to the reader at this time that there are no easy formulas for blasting concrete. Therefore, a small test shot must be made.

UNDERWATER BLASTING

Underwater blasting is necessary for building or deepening harbors and channels, cutting piles, and excavating trenches.

Underwater blasting creates a greater shock wave (which travels through the water), has hydrostatic pressures that must be overcome, requires waterproof explosives, and creates difficult drilling and blasting conditions. Underwater blasting requires reduced burdens, increased subdrilling, and increased powder factors.

Drilling is generally done from a barge with crawler-mounted drills, using 2½- to 6-in (63- to 152-mm) holes. The drill may be taken underwater, but the air exhaust should be vented to the surface.

When done from a barge, the drilling is often done through a sand pipe slightly larger than the gauge of the bit. As the hole is being drilled the sand pipe settles through the loose material to the solid rock to prevent loose rock and mud from falling into the hole. Once drilled the hole may be loaded through the sand pipe, and then the sand pipe can be removed and used for the next hole.

In water 30 to 40 ft (9 to 12 m) deep (approximately one atmosphere of pressure greater than atmospheric pressure at sea level), the powder factor can vary from 1.5 lb/yd^3 (0.89 kg/m^3) to 3 lb/yd^3 (1.78 kg/m^3).

Detonating cord and high-velocity gelatins are common for marine use; however, when delay blasting is desired, a waterproof ammonium nitrate should be used, because it is less sensitive and thus less likely to sympathetically detonate.

Burdens and Spacings

The burdens and spacings for underwater blasting are much less because of the water pressure. The burden will usually range between 0.6 and 0.8 times the burden determination for the same rock type on the surface. If simultaneous firing is planned then the spacing should be the same as the burden. However, if delay firing is used, the spacing will be approximately 1.4 times the burden. The minimum subdrilling should be equal to the burden. This conservatism is because of the high costs involved in having to return to remove high bottom.

Piles

When piles are cut, cartridges of explosives may be tied around the pile, with detonating cord, at the mudline, causing a shearing force when detonated. If 60% gelatin dynamite is used the charge calculation for cutting piles is $0.033\ D^2$, where 0.033 is the powder constant and D is the diameter of the pile (or less dimension for rectangular cuts) in inches. Therefore, a pile 12 in (305 mm) in diameter would require 4.75 lb (2.2 kg) of 60% gelatin dynamite.

Shape Charges In lieu of cartridge explosives for underwater demolition one can use commercial shape charges that are made to the dimension required. With these shape charges, divers merely have to attach the charge to the pile and prime it. These charges are much faster than stick cartridges.

Trenches can be blasted underwater by conventional drilling and blasting techniques or by using precalculated and manufactured shape charges.

BUILDING DEMOLITION

The last few years have given use to a new specialty within blasting: building demolition. The demolition of buildings by the use of explosives has proven to be generally less dangerous than conventional building demolition methods. The obstruction created is short, the dust condition created by the demolition is shorter-lived, and the cost is less.

The principle of blasting down a building basically involves destroying the important supports within the structure so that they fall by virtue of the building's own weight. This is done by careful analysis and planning of where to space the explosives charges. Delayed firing is used to help direct and control the direction of fall. Generally surface charges are not used. However, if they are, they are adequately covered to prevent any flying projectiles from them.

Another consideration in the demolition of a building is whether or not the building lends itself to explosive demolition. That is, the building must be of a minimum height and weight to permit the control of the fall. For example, a single-story wood-frame building would probably not collapse very well under this method. However, a bulldozer could handle the job well.

Building demolition by explosives is extremely hazardous and should not be tried by just any blaster. However, if done by a competent blaster who is experienced at building demolition, it is safer than conventional demolition methods.

TWELVE

UNDERGROUND BLASTING

TUNNELS

The first known industrial use of explosives was the use of black powder to blast in the Saxony mines in the seventeenth century. Tunnels may have been one of the earliest human engineering activities. It is believed that Stone Age men in Europe sank shafts and drove tunnels to find flint, using deer's antlers and flint tools.

Tunnels can be in all sizes and shapes, including round, horseshoe, square, and rectangular. (See Figure 12.1.) The most common tunnel cross section is the horseshoe shape. The reason for this is threefold: first, the roof, or back, forms an arch, which tends to be more self-supporting; second, the flat invert, or floor, gives a work area which is better and easier to move around on with equipment; and third, it is easier to excavate a flat invert.

The theory involved in tunnel drilling and blasting is basically the same as for surface blasting. However, on account of the geometry and the confining underground conditions, some of the blasting problems are increased. To clarify the process, the drill cycle is when the primary activity is drilling the blast holes. The blast cycle is that time spent loading, stemming (if required), tying in, shooting, and waiting for the smoke to clear. And the muck cycle is the time spent excavating the blasted material (muck) from the drifts (the tunnel). These three activities, drill, blast, and muck, are the production cycle.

Whenever possible the support activities are done during one of these activities to maintain a timely advance. Support activities include hanging ventilation pipe, water lines, and air lines; laying track; and setting supports. If the project is large enough, there will be a separate crew, called a "bull gang," to do much of this support work.

If the tunnel is to be lined as it is advanced, then the activities involved in lining the tunnel, shaft, or raise have to be included in the production cycle. For example, sinking a concrete-lined shaft may call for a four-part cycle,

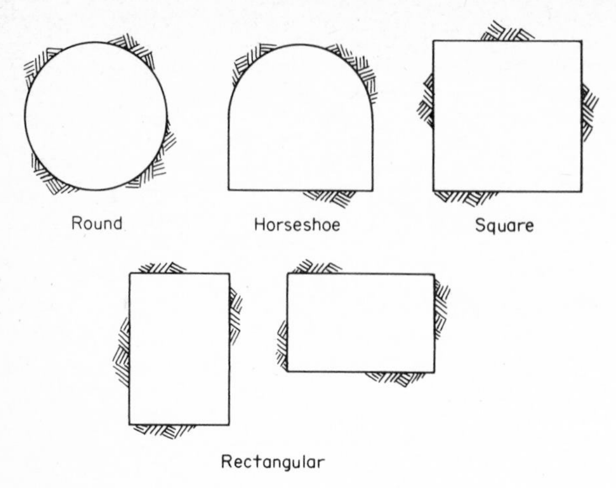

Fig. 12.1 Tunnel shapes.

consisting of (1) drilling, (2) blasting, (3) mucking, and (4) setting forms and placing concrete. The cycle may also be altered so that the sequence becomes drilling, blasting, lowering and setting forms, placing concrete, and mucking. Each project is different, and therefore the best approach will depend on many factors that must be considered at the time.

A tunnel may be advanced by the full-face method, the top-heading–and–bench method, or the pilot-tunnel method.

The Full-Face Method

In the full-face method the tunnel is advanced with a full round; i.e., the entire face is drilled and blasted with each round. Therefore, the heading, or face, advances one complete round after each production cycle of drill, blast, and muck. The full-face method of advance is best suited for small- to medium-sized tunnels.

With the full-face advance most of the holes are drilled at right angles to the face. However, because of conditions, a method must be devised to create a free face for the rock to break toward. As we have seen, a confined blast, such as a tunnel or a shaft round, requires a very large powder factor because of the lack of an area to break to. Methods may be employed to create a free face artificially to enable a reduction of the powder factor and improve breakage.

The methods employed require techniques involving the placement of drill holes. The names of the various types of drill holes in a tunnel round are related to their location (see Figure 12.2). The cut holes are those holes to the center of the face that are primarily responsible for the breaking and movement of the rock in the center of the face. The reliever holes are those holes that help or aid in the breakage of rock by the cut holes; they detonate just

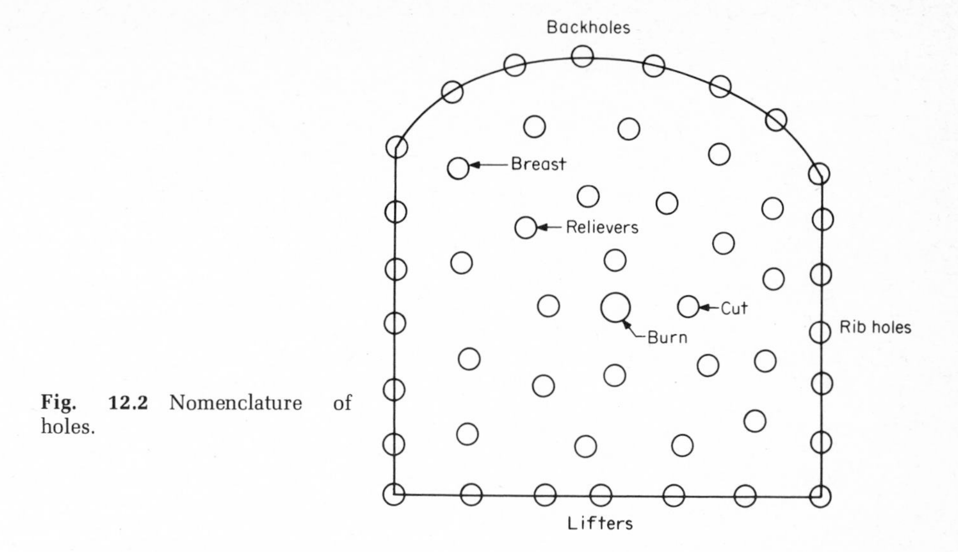

Fig. 12.2 Nomenclature of holes.

after the cut holes and further increase the free face. The trim holes are those holes that are on the perimeter of the round.

The names for the various trim holes are related to their location in the round. The trim holes at the top of the round are called "back holes," the holes along the side of the round are called "rib holes," and the holes at the invert are called "lifter holes."

Breast, or enlarger, holes are those blast holes that are between the reliever holes and the trim holes. These holes are drilled parallel to the axis of the tunnel and generally outnumber the other types of drill holes.

The Burnhole Method The burned-cut method of blasting round first developed as a method utilizing a series of loaded and unloaded straight holes parallel to the direction of advance, or tunnel axis, drilled on 4- to 6-in (102 to 152 mm) spacings (see Figure 12.3). These closely spaced holes offer a plane of weakness to which the loaded burnholes break. In this way, the centerpiece, the area between the holes, becomes a cavity for the rest of the round (the holes surrounding the burn cut) to break toward when it fires after the burnholes. Also, some holes are generally left empty, because if the rock fractures easily the cylinder may have so much rock moving into it that it

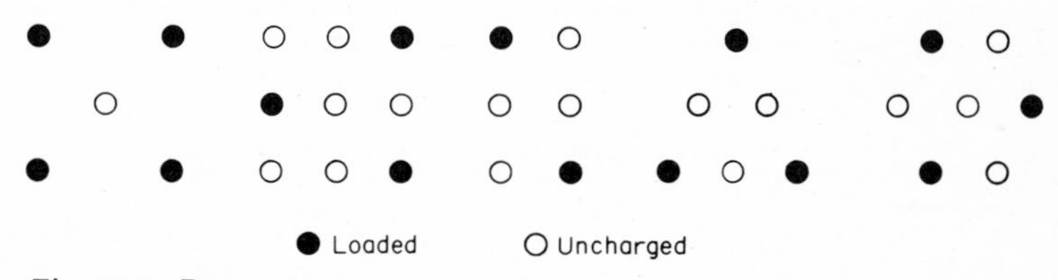

Fig. 12.3 Burn cuts.

clogs, causing the broken rock to become bound against itself, restricting the movement of other rock. (See Figure 12.4.)

The burn cut can also be accomplished by drilling one larger borehole to serve as the burnhole. This drilled cavity serves as a free face for the surrounding holes to break toward. This method is advantageous to the parallel burnholes because the advance is not as restricted as with small burnholes. In other words, the advance may be longer with a larger-diameter burnhole.

For drilling with hand-held drills, such as jacklegs, suppliers sell specially adapted larger bits for burnholes. When drilling 1½- to 2-in (38- to 51-mm) drill holes, one can drill a 3-in (76-mm) burnhole with the same type of drill steel. With larger, jumbo drills, burn cut holes can be 8 in (203 mm) in diameter.

The Coromant Cut: The coromant cut is provided by drilling one large hole as a burnhole and then, by using a special template, drilling another large hole right beside it. This burn cut ends up looking like a figure 8. (See Figure 12.5.) The advantage of this type of burn cut is that the area of the free face is greatly increased.

When the various burn-cut holes are being drilled, care must be taken to drill the holes straight. With the spacing of the loaded holes so close, the chance of sympathetic detonation is quite high (sympathetic detonation is the detonation of one hole by a blast in an adjacent hole). Also, if the holes are drilled inaccurately, the chance of having bootlegs—of not having the rock break the entire length of the borehole—is greater.

Advantages of Burn Cuts: The burn cut offers several advantages in small- and medium-sized headings:

1. Because of the direction of throw with a burn cut there tends to be less flyrock than with the V or wedge cut.
2. The burn cut is drilled straight, permitting more drills at the heading. By

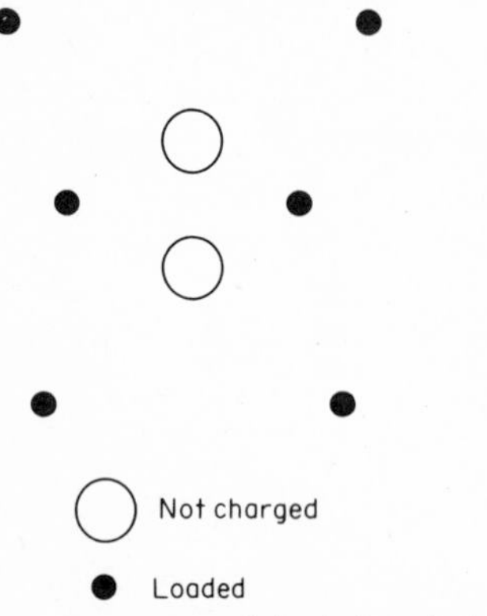

Fig. 12.4 Burn cut with large burnhole.

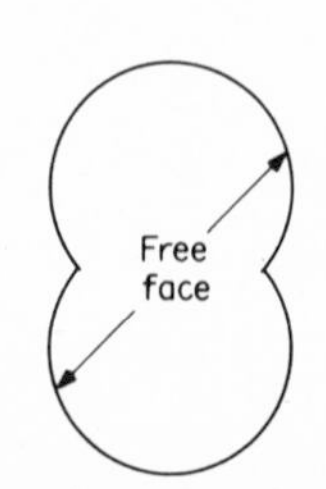

Fig. 12.5 Coromant cut.

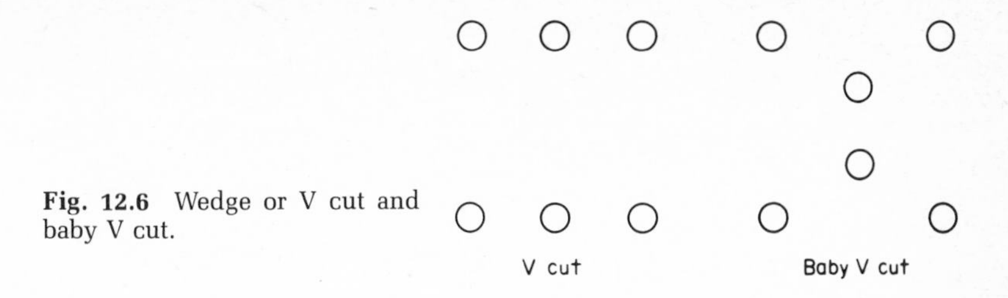

Fig. 12.6 Wedge or V cut and baby V cut.

the same token a wedge cut restricts advancement because of the inability to get greater depth and a proper angle.

3. One can use one total length of steel for all holes.
4. The geometry of hole placement is less critical for the burn cut than for the angle cut.

Angle Cuts Angle cuts are those cuts that have holes drilled at an angle to the face. The wedge cut, also called "V" or "plow," is drilled symmetrically at approximately 60°. Because of the geometry of breakage the wedge cut requires fewer holes and a lower powder factor than the burn cut. (See Figure 12.6.)

The Wedge Cut: The wedge cut is particularly well suited for well-laminated or -fissured rock. It is generally used on wide tunnel faces or in underground rooms where the holes can be aligned with the axis of the tunnel.

A problem with the wedge cut is that the rock from the wedge, being somewhat larger than the other fragments from the blast, can cause considerable damage, because, as it is catapulted from the face, it has enough mass to inflict damage to steel or timber sets (supports). A method to avoid this is to place a small, shallow borehole (a buster hole) in the center of the wedge to blast it into smaller fragments as the face is detonated.

The wedge cut will generally pull an 8- to 12-ft (2.4- to 3.6-m) round; however, the drilling accuracy is quite important. The wedge cut can be made with two or three rows of holes using the V, or wedge, at any angle to the center line of the face. The V can be to the horizontal or vertical, depending upon which direction allows the greatest angle.

The Fan Cut: The fan cut is approximately one-half of a wedge cut. Because of the angle of drilling, which restricts drill space, it is not used often. It is usually limited to pulling a 4- to 5-ft (1.2- to 1.5-m) round and is used only where one drill is used. All the holes are drilled at different angles from one location near either rib. (See Figure 12.7.)

The Draw-Hammer Cut: The draw-hammer cut is another modified V cut in which the holes are drilled away from the center of the face. It is best suited in small tunnels and raises, where there is not much room for drilling. (See Figure 12.8.)

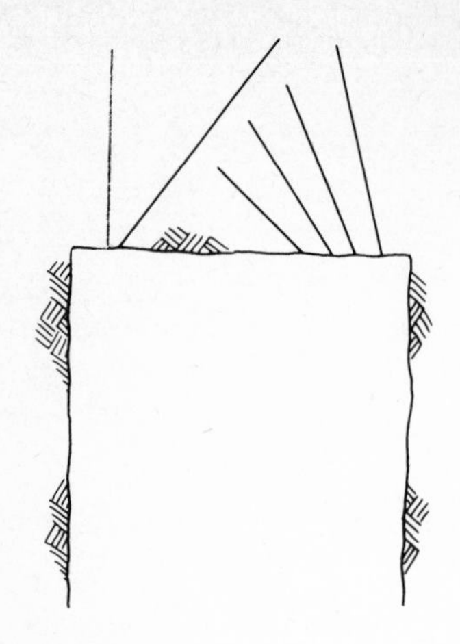

Fig. 12.7 Fan cut.

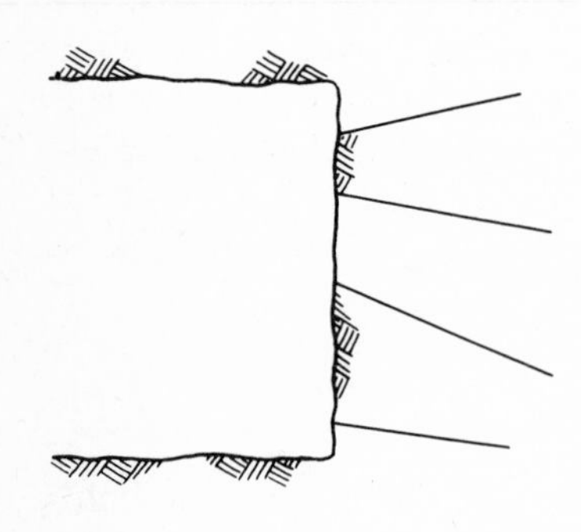

Fig. 12.8 Drift round using bottom draw.

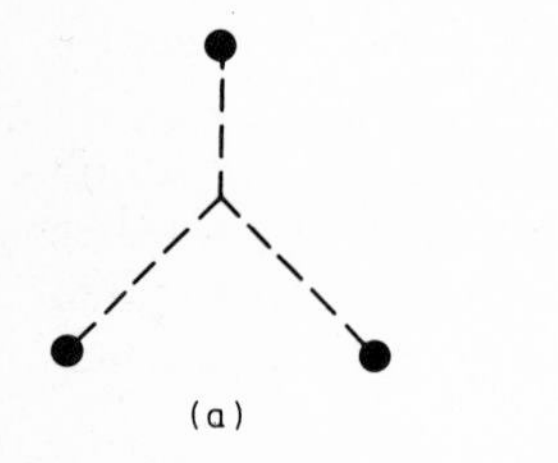

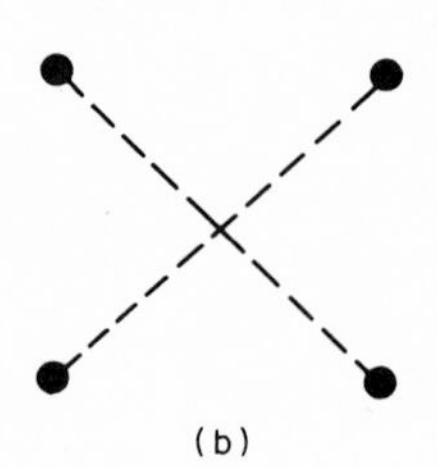

Fig. 12.9 (*a*) Three-hole pyramid for medium-hard cuts; (*b*) four-hole pyramid for hard rock.

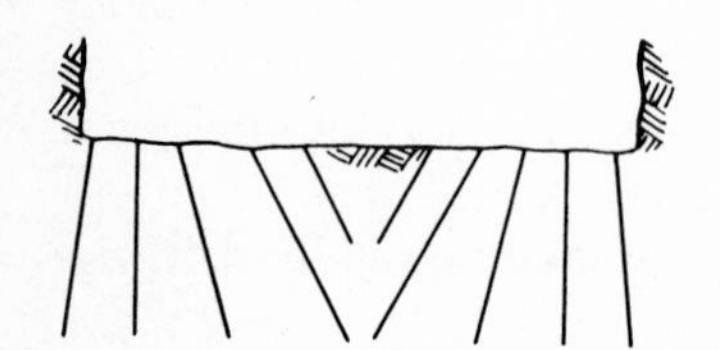

Fig. 12.10 Baby pyramid cut in circular shaft.

The Pyramid Cut: The pyramid cut is similar to the wedge cut, only it has three or four sides. The pyramid cut is used primarily in sinking circular shafts. (See Figure 12.9.)

The pyramid cut holes should be deeper than the other holes to provide a sump when drilling the next round. Because of the size of the round, another pyramid cut, called a "baby pyramid," is required within the perimeter of the larger pyramid cut to aid in the blast geometry and reduce the size of the pyramid wedge rock fragment. (See Figure 12.10.)

Drill Patterns for Tunnel Headings

The drill pattern selected is determined by the size of the face, the type of material being drilled, and the blasting geometry. Each drilling situation varies, so that to give a hard and fast rule for drill hole spacing is impractical. However, rules of thumb for space determination can be made to give an estimate. By no means should these rules of thumb be put into practice; the hole spacing should be determined only by an experienced, qualified individual. These techniques we are about to discuss are for demonstration purposes only; they are empirical formulas and should not be taken literally.

In some cases with a burn cut, the burden in feet is equal to the numerical value in inches of the hole diameter. That is, a hole diameter of 1½ in (38 mm) would require a burden of 1½ ft (0.45 m), or a hole diameter of 2 in (51 mm) would require a burden of 2 ft (0.6 m). The spacing would be equal to 1.4 times the burden; therefore, a borehole diameter of 1¾ in (44 mm) would require a burden of 21 in (0.5 m) and a spacing of 29 in (0.73 m). These values are for medium-hard to hard rock. For softer rock the values assigned will make the burden and spacing dimensions larger. It is possible in soft rock for the burden to be 1.2 times the value in hard rock and the spacing to be again 1.4 times the burden. Therefore, a 1¾-in (44-mm) diameter borehole would have a burden of 25 in (0.63 m) and a spacing of 35 in (0.88 m). (See Figure 12.11.)

To reiterate, these values, or rules of thumb, are a generalization, and by no means should they be construed as exacting formulas. They were used to demonstrate how the burden and spacing may be determined in the stated example and should not be considered an accurate method of drill pattern design.

The drilling procedure is fairly standard whether it is on the surface, in a tunnel, or in a mine. Generally, there should be as many drills at the heading as practical. Before drilling starts, the face must be thoroughly inspected for explosives left undetonated from the previous round. All drill holes in bootlegs must be cleaned out with water to make sure no explosives are left, and the remaining drill hole in a bootleg must never be used to collar or start a new hole. Since alignment is so critical, it is important to drill all the holes to the proper depth and alignment. A method that can be helpful to maintain hole alignment is to place a loading pole in the first hole drilled, leaving a

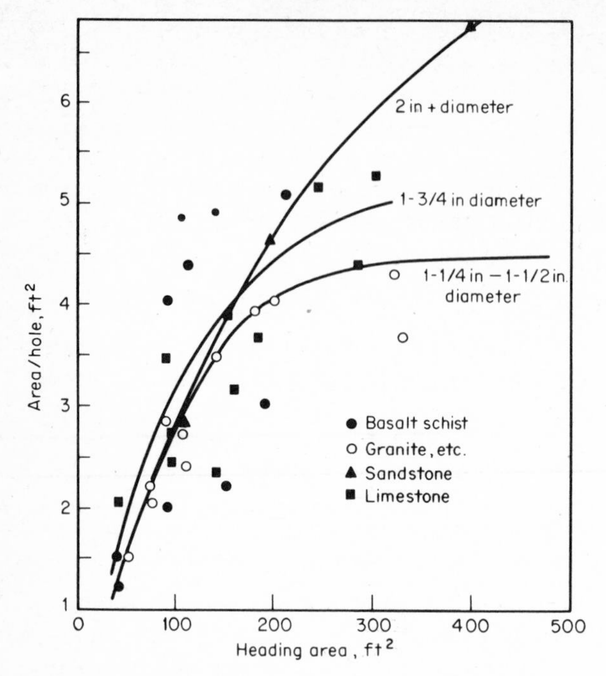

Fig. 12.11 Relationship between the hole diameters and cross-sectional tunnel area. (*E. I. du Pont de Nemours and Co., Inc.*)

few feet of it extending out of the hole. This provides a reference line to which the driller can align the other holes. (See Figure 12.12.)

Loading

The amount of explosives used is, as with surface calculations, considered together with the amount of hole left between the explosives and the collar, or the stemming. A theoretical method for calculating the charge is to make the amount held for stemming equal to 0.7 times the burden. That is, if the burden is 1½ ft (0.45 m), then the amount left for stemming is:

$$\begin{aligned}\text{Stemming} &= 0.7B \\ &= 0.7 \times 18 \text{ in} \\ &= 12.6 \text{ in}\end{aligned}$$

We'll round this to 12 in. Again, as with the drilling, this calculation is for demonstration purposes only.

To determine the powder factor, see Figure 12.13.

When the face is loaded, a wooden loading pole, longer than the hole is deep, should be used to check the holes for obstructions before the charges are loaded and tamped in the hole. The loading should be the responsibility

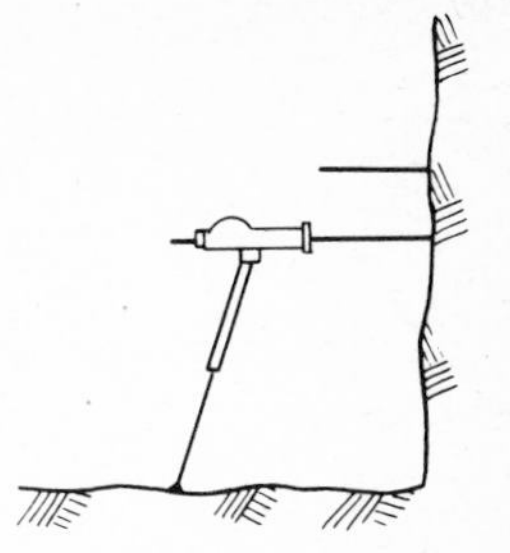

Fig. 12.12 Using loading pole for alignment.

of one person, who watches and supervises the loading cycle. Once the shot is ready to be tied in, all tunnel personnel should be directed from the face and the tie-in should be done by a minimal number of people. The tie-in from the connecting wire to the shooting line should be far enough from the face to protect the shooting line from damage by flying rock and to protect the person who ties the shooting line to the connecting wire, in case of accidental detonation by stray currents in the shooting line.

Delay Caps To obtain adequate breakage, delay caps must be used. Keep in mind the geometry of the breakage, and the delay sequence will be easier to keep straight. That is, the holes involved in the wedge or burn cut must be the first detonated, creating a free face. Then the holes surrounding these holes will detonate in sequence until finally the trim holes detonate. The sequence in which the trim holes detonate relative to each other determines the shape and location of the muck pile. If the lifters (the holes at the invert) fire last, the muck pile will be up against the face. Also, by having one of the

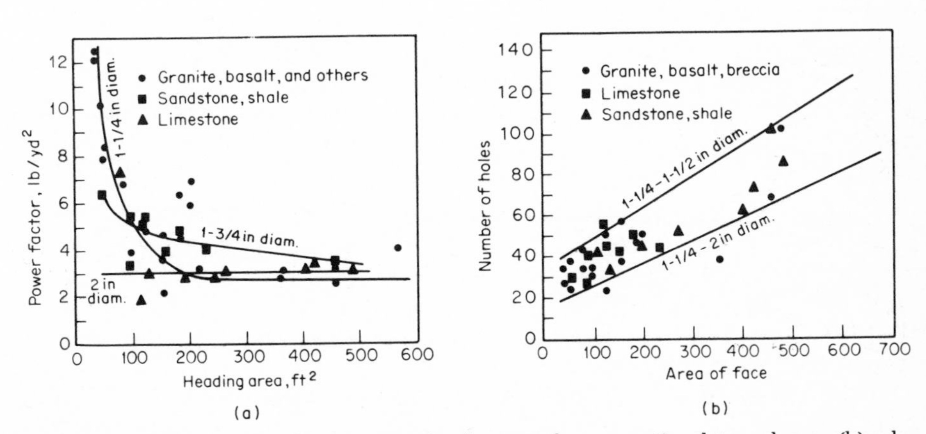

Fig. 12.13 (*a*) Relationship between powder factor and cross-sectional tunnel area; (*b*) relationship between number of holes and cross-sectional area. (*E. I. du Pont de Nemours and Co., Inc.*)

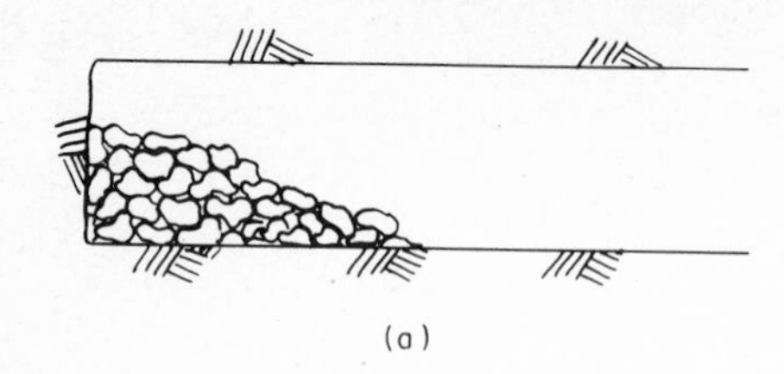

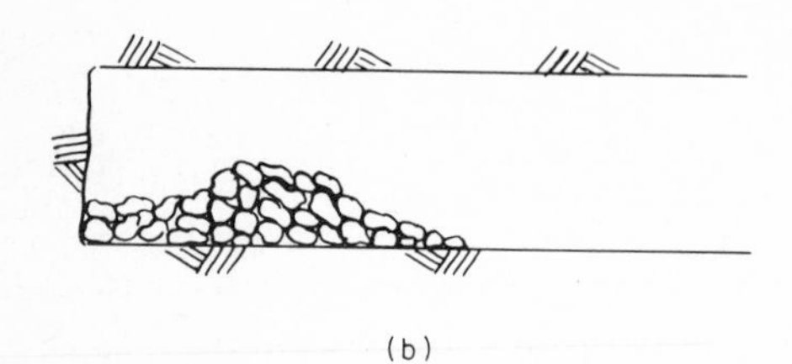

Fig. 12.14 Placement of the muck pile: (*a*) back holes fired last and (*b*) lifters fired last.

ribs fire last you can have the muck pile favor one side of the tunnel. (See Figure 12.14.)

The time period between delays varies according to the burden, the type of rock, the geometry, and the amount of explosives. However, the general range is from 50 to 500 ms, with the larger delays used for more difficult breakage geometry.

As discussed in Chapter 2, explosives used underground must be in fume class 1, and when used in coal mines the explosives must be permissibles. Probably one of the best kinds of explosive for underground blasting, when the breakage geometry permits, is a water gel. The reason for this is its "nonheadache" character. The nitroglycerin headache problem is magnified underground by the decrease in ventilation. Sometimes all that is required to obtain a powder headache is to walk to the heading. Even with adequate ventilation the nitroglycerin fumes will hang in the muck pile.

The choice of explosives is still dictated by the breakage geometry. However, if water gels will do the job, use them.

Advance Heading

In advance heading the tunnel is not driven full-face; that is, one portion is driven in advance of the rest of the face.

Top Heading The most common advance heading method is top heading. The major reason for going with an advance heading instead of full-face is that the roof might be too wide or not stable enough to support itself. When top heading is used the roof support can begin immediately. In other words, before the hole span is exposed, one can have rock bolts, timbers, shotcrete, or any combination of roof supports to enable the span to be increased.

Generally, the top heading is driven first, and the remaining bench is either drilled out from the top of the bench with vertical holes or drilled from the invert with horizontal holes. Drilling from the bench creates the need for subdrilling and thereby causes a rough invert. However, it often allows more drills, because there is more working area, and therefore it is often faster.

Although drilling from the invert may be slower, sometimes the extra time is compensated for because the side trimming is easier, having fewer "tights," which make the invert hard to clean. If it is cleaned to hard rock and concreted, it will require less concrete.

The top heading and bench may be driven simultaneously, or the top heading may be completed and the bench then removed. The decision whether to drive them simultaneously is dependent on:

1. The length of the tunnel
2. The muck disposal
3. The type of lining
4. The comparative cost

The Pilot-Tunnel Method Another method of partial advance is a pilot tunnel. A smaller, pilot tunnel is driven down the centerline of the larger tunnel, generally the entire distance or from shaft to shaft; then the tunnel is enlarged using the pilot tunnel as a large burnhole. Generally the enlargement is drilled from the pilot tunnel with holes perpendicular to the tunnel axis. However, when overbreaking or tights are a problem, it may be more desirable to drill holes parallel to the tunnel axis from the face.

The equipment used for drilling tunnel rounds is generally either jackleg-mounted or jumbo-mounted drills. Jackleg drills (see Figure 12.15) are small percussion drills rotated by pawls and ratchet with a piston size ranging from 2½ to 3 in. Used for drilling small-diameter holes (1½ to 2 in), this drill is used in small drifts. Because of the height limits of the driller, the back holes in a tunnel over 6 ft (1.82 m) high are drilled from upon the muck

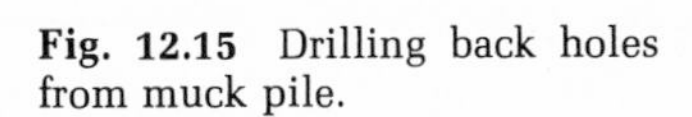

Fig. 12.15 Drilling back holes from muck pile.

pile, which provides the extra height required. Therefore, the back holes are often drilled first and during the muck cycle.

A drill jumbo is a platform or mobile machine, mounted either on a rail, on rubber tires, or on crawlers. The drill jumbo has two or more drills that can drill holes of greater diameter than a jacking drill because of their greater size. Generally a jumbo requires one driller per drill and possibly one helper for every two drills.

SHAFTS

Drilling and blasting for shaft sinking is very similar to driving a tunnel. Therefore, in discussing shafts only the differences will be discussed.

The cuts used for shaft rounds are like tunnel rounds. However, for round shafts the most desirable drill cut is the four-sided pyramid.

Shaft rounds are generally drilled with a shaft jumbo. The shaft jumbo, for a circular shaft, is generally at least a four-boom jumbo. This permits each driller to drill similar quadrants. The shaft drill hole is generally a 2-in (51-mm) hole or larger, with a depth of round approximately one-half the shaft diameter.

Powder Factor

The powder factor for circular shafts is generally quite high because of the mucking difficulty. To muck out a shaft requires either a mucking machine to excavate the muck that it is traveling on, or clamshell-type equipment that is attached to the shaft walls. Either way of mucking requires greater fragmentation of the rock than in driving a tunnel. Shaft blasting generally requires a powder factor of 2 to 6 lb/yd^3 (0.91 to 3.56 kg/m^3).

Sump Cut

For sinking rectangular shafts the sump cut is probably the most popular. The sump cut is particularly well suited for hand drilling and hand mucking. The shaft is sunk one half at a time, alternating between sides so that the side which requires drilling is the higher side and is out of the water. (See Figure 12.16.)

The sump cut is advantageous in that one has a constant free face with a sump for drainage. Also, because it is a bench-type cut, the need for accuracy

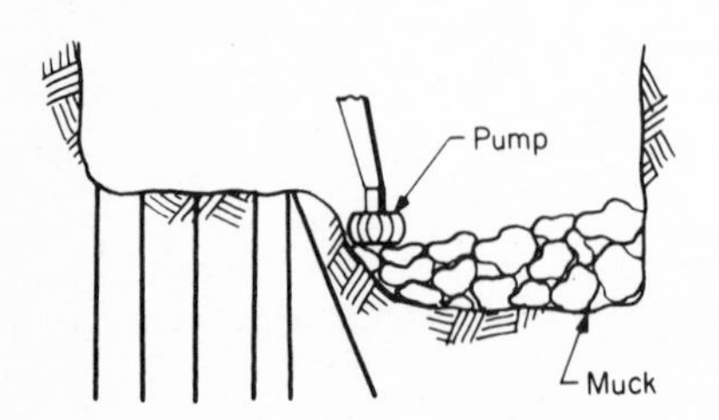

Fig. 12.16 Sump cut.

Fig. 12.17 Nonel system for shafts. (*Ensign Bickford Co.*)

and high powder factors is reduced, resulting in less flyrock. Drilling on a dry bench means that the cap leg wires can be tied together out of the water, decreasing the possibility of misfires.

RAISES

Raises are excavations headed upward from a horizontal drift, or tunnel, either for the purpose of developing ore pockets or as a conveyance for handling materials or equipment. Raises are very hazardous to construct because the driller has to drill off a platform constructed within the raise. After the blast has been fired, the miner must enter the raise, taking the chance of being hit by spalling rock if there has been inadequate breakage. Therefore, it is imperative, not only from a production standpoint but also for safety, that the round be drilled accurately for adequate breakage. However, if there is too much throw the staging can be damaged and therefore unsafe.

Muck removal is not a problem, because gravity takes care of it: the muck falls from the face area to the bottom of the raise.

Raises are usually drilled with just one jackleg, because the room is quite limited, and the cut holes are generally drilled 6 in (152 mm) farther than the advance to reduce the chance of bootlegs. Burn cuts are the most common in smaller raises; however, as the size increases the tendency is toward wedge cuts.

For very large raises there are climbing raise jumbos that climb the raise on their own monorail fastened to the wall of the raise. This permits the raise to be drilled and loaded from a protected deck, which is then removed from the raise for protection from the blast.

THE NONEL SYSTEM

In recent years the development of nonelectric delay firing systems has added greater safety to underground blasting. These methods are gaining wide popularity and offer delay blasting without the risk of blasting electrically underground.

The Nonel system, as discussed in Chapter 5, permits the loading and tying in of blast holes underground without fear of accidental detonation from stray electric currents. Figure 12.17 shows some of the advantages of the Nonel system.

THIRTEEN

ESTIMATING

IMPORTANCE OF ESTIMATING

In a contracting or engineering situation, technical expertise is often not enough. A contractor engaged in drilling and blasting not only has to be a technician but also has to possess the skills and abilities to determine the costs involved in performing the task. In other words, it is as important for the contractor to be able to estimate as it is to blast. If a contractor cannot estimate properly either the bid will be so high that he or she will never be awarded any contracts or the bid will be so low that after one or two projects he or she will be investigating other areas of employment or business.

These words of caution to the contractor do not relieve the engineer of the responsibility of understanding estimating. Generally before a job is released for bid an estimate has to be done by the engineer so that the client (the one paying the bills) can determine whether to accept or reject the proposed project or to aid the client in planning for funding. Often, when the job is in progress blasting may be required as an extra to meet a changed condition. Then it is the responsibility of the resident engineer to determine a fair price.

This chapter will familiarize the reader with all the various kinds of considerations for preparing an estimate for drilling and blasting rock. (To be complete, the chapter is directed toward the reader who has no prior knowledge about preparing an estimate.) There is a discussion of quantity takeoffs, specifications, equipment selection, costs, production, and strategy. All of this must be well understood before a competent estimate can be produced. What also has to be understood is exactly what has to be done. It is reasonable to assume that the job requires drilling and blasting, and that the estimator has to know how much; but that "how much" doesn't mean just the volumetric quantity (how many cubic yards), but also many details, such as the depth of the cuts, the amount of line drilling, the minimum permissible breakage, and the amount of drilling required.

QUANTITY TAKEOFF

The first step in the estimating is to do a quantity takeoff to determine the total amount of work to be done. There are several categories of quantity for the typical drilling and blasting job. The primary quantity needed is the volumetric quantity, or the cubic dimensions of rock to be blasted. This quantity has to be broken down, or categorized, into its components. Typical categories are cubic yards of open blasting (massive rock cuts that do not require mats); mat work (blasting that, on account of its proximity to structures, people, or other things to which blasting could present danger, must be covered by blast mats or like devices to control or limit flyrock); and trench blasting, which may be segregated because, as a general rule, it is more expensive per unit than open blasting (often there is a separate bid item for trench blasting because of this realization).

As an example of the three categories of quantity thus far discussed, a project with 65,400 yd^3 (50,000 m^3) may be broken down into 39,240 yd^3 (30,000 m^3) of open blasting (often referred to as "class A rock"), 6540 yd^3 (5000 m^3) of mat work, and 19,620 yd^3 (15,000 m^3) of trench blasting (often referred to as "class B rock").

Organizing the Quantity Takeoff

Generally the quantity takeoff should be segregated by either bid items or different unit costs, whichever creates more categories. Using unit costs is strictly an empirical judgment; it is accepted that mat work generally costs more per unit than open blasting, as does trench blasting. Also, one may want to categorize the quantities for the different depths of cut; for example, a 6-ft (1.8-m) depth of cut generally creates a greater unit cost than a 20-ft (6.1-m) cut. If the quantity breakdown has more separate groups than bid items, the costs are still computed by the breakdown and then totaled into the bid item categories.

Total Quantity

When the quantity takeoff is done, the total quantity must also be taken into account, i.e., the difference between neat line and actual quantity. For example, the foundation for an industrial building has the dimensions 164 ft by 328 ft (50 m × 100 m) and a cut varying from 6 to 18 ft (1.8 to 5.9 m) and averaging 10 ft (3 m) in depth. Our first instinct would be to compute the quantity by multiplying the linear dimensions to get the volume. Doing this we obtain 19,922 yd^3 or 15,000 m^3. This figure is proper to use as the pay quantity; however, using this quantity to determine the amount of work to be done is quite erroneous and could lead to a very costly error. Referring to Figure 13.1, let us again determine the quantity.

First, as you will note, 164 ft × 328 ft is the foundation area. But the construction of the foundation is going to require additional space. For the sake of this example let us assume that carpenters need an additional 1½ ft (0.46 m) on all sides to be able to construct a footing and erect wall forms. This increases the quantity to 167 ft × 331 ft × 10 ft, or 20,473 yd^3, an increase of 551 yd^3 (421 m^3). However, by now we are well aware that rock does not grace us by breaking in a straight line following the outside row of boreholes. Therefore, we must also take into consideration the angle of breakage. This again requires an increase in the dimensions of the foundation plan. If we are using a spacing of 6 ft (1.8 m) and moving the outside row of holes one-half the spacing distance beyond the desired break line, we move our perimeter rows of hole out another 3 ft (0.9 m). This again changes the dimensions to 173 ft × 337 ft (52.7 m × 102.7 m).

Once the additional work has been totaled for the horizontal dimensions, that for the vertical must be considered: the subdrilling. We have learned earlier that to ensure that all the rock is broken to below the required elevation, we must subdrill (drill below the required elevation). For this example, let us use half the spacing for our subdrilling distance. Therefore, the average depth of cut goes from 10 ft (3 m) to 13 ft (3.96 m).

Computing the final or actual quantity, the dimensions are 173 × 337 ft × 13 ft, or 28,071 yd^3 (21,463 m^3). This is an increase of 8149 yd^3 (6231 m^3), or 41%.

To conclude, we have taken the pay quantity of 19,922 yd^3 and demonstrated how the actual quantity may greatly exceed the pay quantity. In this example it was 41% more; however this can vary greatly depending upon the original dimensions. If the foundation had been 650 ft (198 m) by 1300 ft (396 m) we still would have added only 3 ft to each of the horizontal dimensions; therefore, the percentage of increase would have been much less. The most critical change in the example was the depth of cut. The depth, to allow for subdrilling, increased from 10 ft (3 m) to 13 ft (3.98 m), which was an

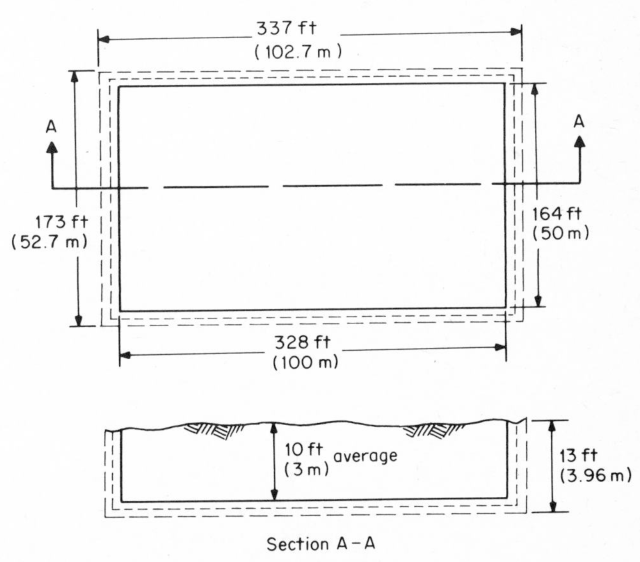

Fig. 13.1 Foundation layout demonstrating the difference between the actual and the pay quantities.

Date ________ Job Location ________ Job Number ________

Bid item	Dimension Length	Dimension Width	Dimension Depth	Bid quanity	Over drill Length	Over drill Width	Over drill Height	Actual quantity	Comments

Fig. 13.2 Quantity takeoff sheet.

increase of one-third. However, had the original depth of cut been 16 ft (4.8 m) and had we added 3 ft (0.91 m) for subdrilling, the net increase would have been approximately one-fifth, or 20%. Therefore, it is important to check the relative quantities for each project. Never assume a standard percentage; always check and record the actual percentage on the quantity takeoff sheet. (See Figure 13.2.)

Now that we have discussed the quantity takeoff let us analyze the effect that the specifications can have on the estimate.

SPECIFICATIONS

The specifications are those parts of the bid package, or contract, that supplement the drawings with the complete technical requirements of the job and generally take precedence over the drawings. It is essential that the specifications be read (even studied) before a bid is submitted. The specifications will often explain what the blaster may and may not do in the performance of a task. They will also establish certain performance criteria within which the blaster must produce. For example, the specification may require that more of the breakage may not exceed 1 yd^3 (0.765 m^3) in size. This has a definite bearing on the blasting, from preblast design to the amount of secondary blasting. Also the specification will probably require blasting mats. Generally the specifications require mats but the interpretation is very loose as to how mats are required. The ambiguity of most blasting specifications is generally for some pragmatic reasons. Blasting specifications, especially with regard to mats, will often require mats but leave it to the discretion of the engineer when they will be used. The reason for this is that the specification must require mats but the writers realize that they may not be needed for the entire job. The more blasting that can be done without mats, the lower the cost. In this way the engineer responsible for the specification establishes the need for mats, avoiding a future extra work claim, and (in many areas) keeps the specification in conformance with local laws regulating blasting.

Common Errors

A common error made by blasting contractors regarding the prebid analysis of the actual portion of the job that requires mats is using what they think will be necessary without confirming it with the engineer. The blasting contractor who contacts the engineer for an interpretation of the specification is ahead of the game. The engineer may be quite knowledgeable in blasting operations and therefore may have very fair interpretations; or the engineer's interpretation may be more conservative because of a lack of understanding (or more knowledge than the contractor has), thus increasing the blasting costs.

Another pitfall that the blaster has to be aware of is the restrictions on blasting around buildings and other structures. This can vary both in distance and blast design constraints. A specification may explicitly prohibit blasting within a certain distance of structures, or it may impose blasting restrictions in the form of limiting the weight per delay for different proximities to structures or for the entire job. The specifications may limit the depth of cut, necessitating benching. For example, if the total cut is 100 ft (30.5 m) and the specification limits the blasting to 33 ft (10 m), benching will be required. The cut will have to be done with a minimum of three lifts or benches. (See Figure 13.3.)

For blasting around recently placed concrete, the specification often prohibits blasting within 7 days following the pour date. This may be more of a planning than an estimating problem; however, it should be considered.

Finally, the specification may require line drilling within certain distances of structures. Also the specification generally establishes the line-drilling pattern. For example, the specification may require that the web (the

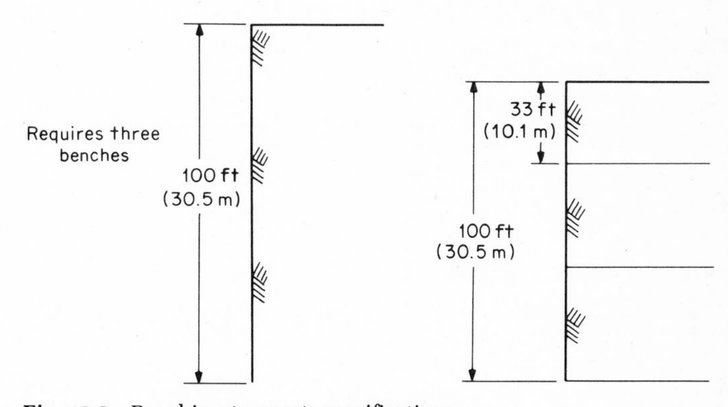

Fig. 13.3 Benching to meet specifications.

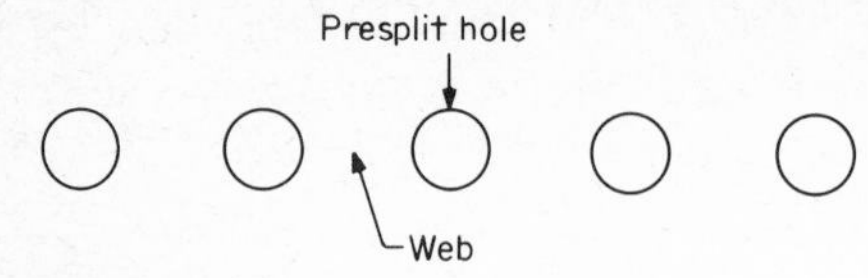

Fig. 13.4 Presplitting to meet specifications.

distance between line-drilling holes) be equal to the hole size. Thus, a 2 ¾-in (70-mm) line-drilling hole would require a 2 ¾-in web. (See Figure 13.4.)

COSTS

After determining quantities, categories of quantities, and specification requirements and restrictions, the various costs involved must be determined and analyzed.

Direct Costs

There are two basic kinds of costs related to all estimates: direct and indirect. Direct costs are all those that can be traced directly to the unit, including labor, explosives, equipment, and other tools and materials consumed on the project. Indirect costs are primarily overhead costs (such as office trailers and secretaries), which may be split into job overhead and company overhead.

Explosives Costs Before the cost of explosives on the job can be calculated, the preliminary blast design will have to have been completed in order to determine the powder factor and the number of blasting caps required. If it has been calculated that the average powder factor for the job is 1 lb/yd^3 (0.589 kg/m^3) and the total quantity is 65,400 yd^3 (50,000 m^3), then the total explosives consumption will be 65,400 lb (29,665 kg). However, the total quantity of explosives has to be analyzed and categorized according to cost. For example, if it were determined that of the 65,400 lb (29,665 kg) of explosives, 5400 lb (2449 kg) is gelatin dynamite, 10,000 lb (4536 kg) is slurry, and the remaining 50,000 lb (22,680 kg) is ammonium nitrate, the cost would look something like the following:

Quantity	Explosive	Unit price	Cost
5,400 lb (2,449 kg)	gelatin dynamite	$0.60/lb	$ 3,240
10,000 lb (4,536 kg)	slurry (water gel)	$0.55/lb	5,500
50,000 lb (22,680 kg)	ammonium nitrate	$0.12/lb	6,000
65,400 lb (29,665 kg)			$14,740

However if the explosive were not categorized by costs the estimate could be unrealistic, thereby jeopardizing the bid.

To determine the number of blasting caps required, it is basically a matter of calculating the number of blast holes required and multiplying the number by a factor (generally 1.05).

To calculate the number of blast holes, one first determines the drilling area (on the horizontal) in square feet and divides through by the area of the drill pattern. Returning to the example in Figure 13.1,

$$173 \text{ ft} \times 337 \text{ ft} = 58{,}301 \text{ ft}^2$$
$$\text{Drill pattern} = 6 \text{ ft} \times 6 \text{ ft} = 36 \text{ ft}^2$$
$$\frac{58{,}301}{36} = 1619 \text{ drill holes}$$
$$1619 \times 1.05 = 1700 \text{ blasting caps}$$

Equipment Costs In determining equipment costs there are several factors to take into consideration before one can begin to estimate. However, since the process of selecting equipment was covered in Chapters 3 and 4, we will arbitrarily select the equipment for this discussion on estimating.

The unit equipment cost is in part a function of the hourly costs relative to the hourly production.

$$\text{Operation cost per unit of equipment} = \frac{\text{Hourly cost of equipment operation}}{\text{Hourly production}}$$

This formula holds true for most unit price production. With drilling and blasting, this formula has to be further adapted to apply for the pricing unit. In other words, the formula will give the linear unit price for drilling, which then must be converted to the volumetric. For example, if it is calculated that the price for drilling is \$0.90/ft (\$2.95/m), the volumetric price on a drill pattern of 6.5 ft × 10 ft (2 m × 3 m) is equal to

$$\frac{6.5 \text{ ft} \times 10 \text{ ft}}{27 \text{ ft}^3/\text{yd}^3} = \frac{2.4 \text{ yd}^3}{\text{ft}}$$

which yields \$0.38/yd^3 (2 m × 3 m × 1 m = 6 m^3/linear unit, or \$0.49/m^3).

Equipment costs can be determined by various methods; however, all methods must include the costs that we will discuss in this chapter. These costs can be broken down into three major categories: cost of ownership, cost of operating, and cost of renting. The easiest of these three categories of cost to determine is cost of renting. Generally, rental cost is a charge on a basic unit of time. That can be anywhere from an hourly charge to a monthly charge. Therefore, this is a known cost and is fairly easy to recognize and define.

Cost of Ownership The cost of ownership is generally the most difficult of equipment costs to determine. It includes the delivered purchase price of equipment, less salvage value, if any, amortized over the selected life of the equipment, plus depreciation, interest, taxes, insurance, and storage.

Delivered purchase price and depreciation are really part of the same calculation. Depreciation is that part of the equipment's life that has passed or has been utilized. For example, if a machine is purchased with an anticipated life of 5 years (yr) based on use of 2000 h/yr, every hour that the machine is used the machine lessens in value one ten-thousandth of the delivered purchase price. In other words, if a machine's delivered purchase price is $100,000 (and we assume no salvage values), then for every hour, the machine's cost is $100,000/10,000 h, or $10/h just to cover depreciation.

The delivered purchase price should include all costs connected to taking delivery of the equipment: the purchase price, extras, freight, assembly costs (if any), and sales tax.

The remainder of the ownership costs can be determined by multiplying the cumulative percentage by the average yearly investment. The average yearly investment is simply

$$\frac{U + 1}{2U} \times C^{*}$$

where U is the estimated life of the machine in years and C is the delivered purchase price. Using the same example, we find

$$\begin{aligned}\text{Average annual investment} &= \frac{5 + 1}{2(5)} \times \$100{,}000 \\ &= \$60{,}000\end{aligned}$$

The cumulative percentage of interest, taxes, and insurance and storage may be computed by using 12% for interest (this will vary from year to year). Taxes will vary according to location; however, for this discussion we will use 2%. Insurance and storage will also vary (again we will use 2%). This gives a combined percentage of 12% + 2% + 2% = 16%. To find the hourly cost of taxes, interest, and insurance and storage multiply the average annual investment by the cumulative percentage and divide by the total hours of use per year. In our example:

$$\frac{16\% \times \$60{,}000}{2000} = \$4.80/\text{h}$$

Operating Costs: Operating costs are those costs that are created by the actual operation of the equipment—repairs, fuel, lubrication, tire replacement and repair (if applicable), and labor.

Repairs are those costs associated with replacing and repairing parts to the equipment. They show a kind of inverse relationship to the depreciation, or ownership cost. Equipment, like automobiles, depreciates quickly when new but more slowly as the machines age. Therefore, the actual hourly cost of ownership falls throughout the life of the machine. The reverse of this is true for repair costs. The hourly repair costs, like the depreciation, vary

* David A. Day, "Construction Equipment Guide," Wiley Interscience Publications, John Wiley and Sons, New York, 1973, p. 39.

throughout the life of the equipment, but the repair costs start out low and increase as the machines age. (See Figure 13.5.)

The amount calculated for hourly repair cost is best determined from historical data based on a percentage of the hourly ownership costs. Lacking firm historical data (maintenance records of similar equipment), one must obtain the information from the manufacturer of the equipment. It will vary according to the type of equipment, the use of the equipment, and the quality of maintenance.

Fuel costs are the hourly consumption of fuel times the cost of fuel. However, one may wish to include the cost and operation of the fuel truck, if one fuels one's own equipment. This truck and operator cost may be included in either the hourly fuel costs or overhead. If the spread of equipment is large enough to warrant one's own fuel truck and operator it may be better to include this cost in the hourly fuel cost.

The fuel consumption will vary according to size and type of engine, use of the equipment, and operating efficiency. The best insight into fuel consumption is historical data (consumption in the past).

The hourly cost of lubrication can vary from 20 to 50% of fuel costs. A servicing schedule is a good indicator: in addition to regular greasing, a schedule based on operating hours is the best way to arrive at servicing costs. For example, if a machine, say a compressor, has all the oil and filters changed at regular, 100-h intervals, then the total cost of this servicing (including labor) can be divided by 100, and this is the hourly cost. However, this cost should be then converted into a percentage of fuel costs. This serves as an allowance for inflation. Lubricating materials, being petroleum products, will generally undergo price increases similar to fuel price increases.

Operator Costs: Labor or operator costs are the total cost per hour for personnel to operate machines. These costs may include the operators' wages, employer-paid taxes, pension, health insurance, vacation, insurance (workmen's compensation), and other costs in the labor agreement. Generally, the costs in addition to wages, often referred to as "burden," can be converted to a percentage of the wage for ease in large computations.

Bits and Steel Another cost associated with drilling and blasting is the cost of bits and steel. This can be computed by the cubic yard (cubic meter) or

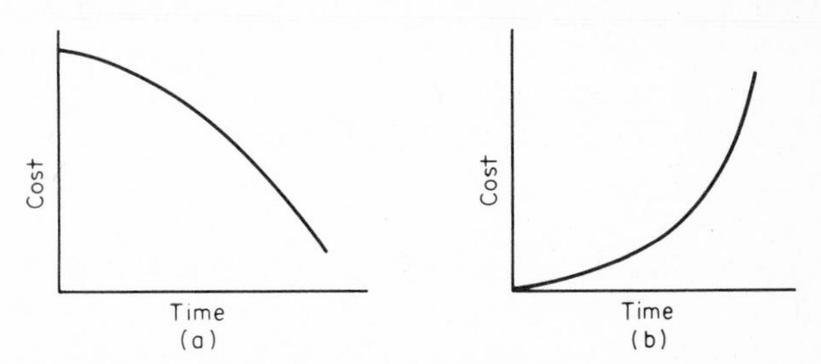

Fig. 13.5 (*a*) Depreciation curve and (*b*) repair curve.

linear foot (linear meter); however, the linear foot is preferred. Again the best source for this information is historical data. This is based on the life or total amount of drilling that the bits and steel demonstrated on past jobs of similar rock characteristics.

Labor Costs Labor requirements and costs vary from job to job according to the physical requirements of the job and the various labor customs or union agreements. Therefore, to determine the labor costs for a particular job it is essential to understand what kind of crew is required. A union job may require a compressor operator, whereas on a nonunion job the driller may start the compressor. Also, to decrease compressor operation costs on a union job one may elect to "bank" the air, i.e., to feed several drills off a manifold connected to fewer, larger compressors. For example, in some union jurisdictions the agreement calls for one compressor operator per two compressors. If the air requirement is 2400 ft^3/m one may elect to use two 1200-ft^3/m compressors and one compressor operator as opposed to four 600-ft^3/m compressors and two compressor operators. Of course, the availability and cost of larger compressors will affect this decision. Also, if the job is spread out and consists of many smaller rock cuts the logistics may not permit banking the air.

Another consideration in sizing the crew is the terrain. On mountainous steep slopes the drillers may require a helper or the blasting crew may require more personnel because of the slow working conditions. To conclude, each job will have its own crew requirements and must be considered in the light of the aforementioned conditions.

Production Production is the key element in any estimate. It is the factor that can make or break a job. It is that unit-per-time figure that is probably the most difficult to determine and certainly the most important. Your accuracy in determining production rates has the greatest single effect on the profitability of the job. If the production is underestimated, the bid price will probably be higher than that of the competition and therefore you will not be awarded the contract. Conversely, if you overestimate the production rates, you will probably lose money on the job. The reason production has such a profound effect relative to cost determination is that production is all-encompassing. For example, suppose on an estimate there is a mistake in the amount of explosives required. Let us assume that the job requires 60,000 kg but there is a transposition error made and the cost is computed on 6000 kg. This means that the explosive costs in the estimate are only 10% of what they will in actuality cost. However, if we assume for the sake of the example that the explosive costs are 5% of the total cost, the net result of our error is that the estimated cost falls 4.5% short of the actual cost. For example, assume we have a project with a cost of $1 million. The explosives cost of this job, at 5%, is $50,000. However, an error was made and the cost was figured at $5000.

That would have made the estimated costs $955,000, a difference of $45,000. This $955,000 is 95.5% of the actual job costs. That is an error of 4.5%.

Now to use the same project and make a similar error on the production rate we can see a considerable difference. The production rate is, in effect, labor costs, equipment rental or ownership costs, and overhead. For the sake of this example let us assume that these costs affected by the production rate amount to 70% of the total job costs. In our job cost of $1 million the production rate accounts for $700,000 of the cost. If an error of the same magnitude is made then the estimated job cost will be computed as follows: A production rate 10 times the actual rate reduces that portion of the estimate from $700,000 to $70,000. The net result is $370,000 for the estimated cost of a job that actually will cost $1 million to perform. This is 53% too low.

It is unlikely that an error of this magnitude would not be eventually caught. Even in a bid situation, a bid this much below the competition's would be questioned. However, as unlikely as this error may seem to the reader, it is indicative of the effect that the production rate does have on the overall job costs.

Indirect Costs

The indirect costs associated with an estimate are generally referred to as "overhead." Overhead can be broken down into two categories: job and company. Job overhead is that portion directly attributed to the job. It may include office trailers, job telephones, the project superintendent, the timekeeper, the superintendent's pickup, and other expenses. "Company overhead" refers to the overhead that is shared in both benefit and cost by all jobs. Examples of company overhead are the home office, secretaries, the equipment (yard and shop), and the president's yacht.

Although often done by percentage, the best method for determining job overhead is to actually calculate the various costs involved, such as the monthly rent for office trailers, pickup, and other such items. However, company overhead is generally determined by percentage. This percentage is generally based on history, within certain limits. For example, if for the last 2 years company overhead has been equal to 50% of the job cost, it would be considered, by most, to be unwise to add 50% of the estimate for company overhead. Most would agree that in this case it would be time well spent determining why the company overhead has been equal to 50%. Generally, company overhead is in the range of 5 to 15%.

Overhead is an important cost to watch and keep under control. Often the growing contractor permits overhead to grow faster than the volume of work warrants. This can cause a good company to "go bust." The destructive thing about excessive overhead is that overhead by its very nature is nonproductive and generally is the hardest cost to reduce. Therefore, when estimating, although competition may necessitate the use of the standard overhead allowances, it is still a good idea to know exactly what all overhead costs are and how much.

The Estimate

Utilizing the techniques or methods discussed in this chapter, let us now assemble a complete estimate. When constructing an estimate it is important to keep it organized; therefore, in addition to quantity takeoff sheets one should develop one's own estimating forms. The forms can be set up for hourly, daily, or total production.

Reading the specification we learn that it is a highway job and all cuts that are in excess of 6 ft (1.8 m) must be presplit. The specification calls for a

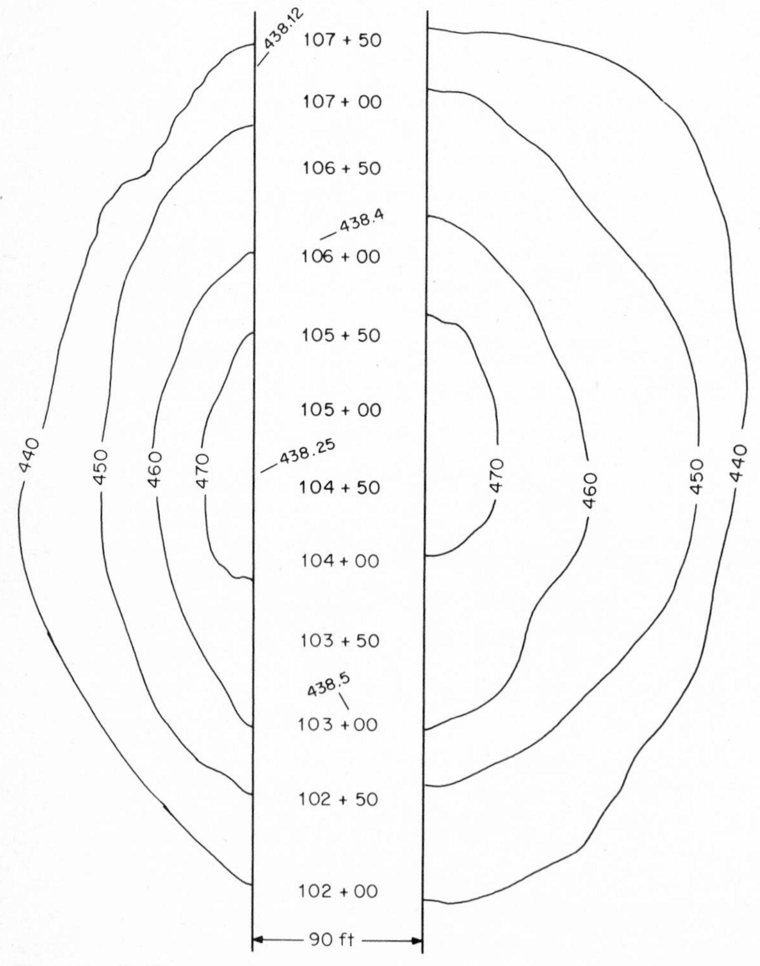

Fig. 13.6 Road cut in plan view.

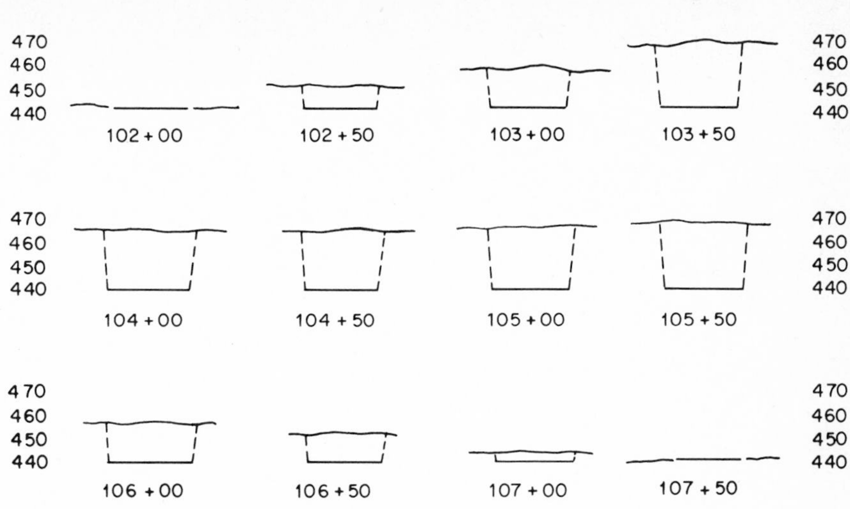

Fig. 13.7 Cross sections of the road cut.

presplit on a slope of ½ : 1 (for every 2 ft of vertical distance there is 1 ft of horizontal distance) with a maximum spacing between presplit holes of 3 ft (0.91 m). On the bid form we see that there is only one pay item to cover drilling and blasting; therefore, all costs will be translated into one unit price.

Takeoff

The second step is to do a quantity takeoff, taking into account the preliminary blast design and all the necessary overdrilling. Using the plans in Figure 13.6 we first must compute the volumetric quantity and the linear drilling required. To facilitate our example we establish certain knowns with the blast design. For the sake of this example, assume:

1. Priming of the presplit is to be done with detonating cord.
2. The powder factor is 1 lb/yd^3 (589 g/m^3).
3. The drill pattern is 6.5 ft × 10 ft (2m × 3m).
4. Subdrilling is 3.25 ft (1 m).
5. Benching is not required.
6. The presplit density is assumed to be 0.45 lb/ft (62 g/m).
7. The presplit slope is ½ : 1.

Figure 13.6 is a plan view of the highway cut with stationing and contours. Figures 13.7*a*, *b*, and *c* are cross sections of the road cut. Table 13.1 shows the average depth of cut for all cross sections.

The best way to compute the volumetric quantity is by the "average end area" method; i.e., determining the average area between each two stations,

TABLE 13.1 Average Depth of Cut

Station	Depth of cut, ft
102 + 00	0
102 + 50	12
103 + 00	20
103 + 50	35
104 + 00	37
104 + 50	35
105 + 00	34
105 + 50	36
106 + 00	25
106 + 50	16
107 + 00	7
107 + 50	0
Total	257

Average depth of cut is 21.4 ft (6.53 m).

then determining the average for all such pairs and multiplying it by the distance between stations. In Figure 13.7, we wish to compute the area of the cross section at 102 + 50. We use the 12-ft average cut, the width of the excavation, and the presplit slope* of ½ : 1.

$$12 \text{ ft} \times 90 \text{ ft} + 12 \text{ ft} \times 6 \text{ ft (the presplit dimension)} = 1152 \text{ ft}^2$$

For the metric equivalent

$$3.66 \text{ m} \times 27.4 \text{ m} + 3.66 \text{ m} \times 1.82 \text{ m} = 106.97 \text{ m}^2$$

Next compute the cross-sectional area for station 103 + 00:

$$20 \text{ ft} \times 90 \text{ ft} + 10 \text{ ft} \times 20 \text{ ft (the presplit dimension)} = 2000 \text{ ft}^2$$

For the metric equivalent,

$$6.09 \text{ m} \times 27.43 \text{ m} + 3.05 \text{ m} \times 6.09 \text{ m} = 185.6 \text{ m}^2$$

By finding the average of these two cross-sectional areas and multiplying the average by the linear distance between the stations, we obtain the volumetric quantity between the two stations. We obtain:

$$\frac{1152 \text{ ft}^2 + 2000 \text{ ft}^2}{2} = 1576 \text{ ft}^2 \text{ average area}$$

$$1576 \text{ ft}^2 \times 50 \text{ ft} = 78{,}800 \text{ ft}^3$$

$$78{,}800 \text{ ft}^3 = 2918.5 \text{ yd}^3$$

Metric equivalent:

$$\frac{107.03 \text{ m}^2 + 185.6 \text{ m}^2}{2} = 146.32 \text{ m}^2$$

$$146.32 \text{ m}^2 \times 15.24 \text{ m} = 2229.9 \text{ m}^3$$

* The two right triangles, one on each side of the cut, created by the ½-to-1 slope of the presplit are added together and treated as a rectangle for purposes of computing the cross-sectional area.

Table 13.2 shows the same calculation for all the cross sections in the highway cut in Figure 13.6 and gives the total volumetric quantity of 50,919 yd^3.

The next quantity to determine is the total linear feet of drilling required. The length of the road cut is 550 ft (167.6 m) and the width is 90 ft (27.4 m). Multiplying these figures gives the horizontal area in square feet (it is not necessary to determine the overbreak, because there is presplitting).

$$550 \text{ ft} \times 90 \text{ ft} = 49{,}500 \text{ ft}^2$$

Metric equivalent:

$$167.6 \text{ m} \times 27.4 \text{ m} = 4592.24 \text{ m}^2$$

Then dividing this figure by the square feet of coverage provided by the drill pattern gives the number of holes required in the cut. For a drill pattern

TABLE 13.2 Cross Sections and Volume of Rock

Station	Cross section, ft^2	Average end area (average cross section between stations), ft^2	Distance between stations, ft	Quantity, ft^3
102 + 00	0			
		576	50	28,800
102 + 50	1152			
		1576	50	78,800
103 + 00	2000			
		3324	50	166,175
103 + 50	3763			
		3889	50	194,425
104 + 00	4014			
		3889	50	194,425
104 + 50	3716			
		3726	50	186,300
105 + 00	3689			
		3789	50	189,425
105 + 50	3888			
		3225	50	161,263
106 + 00	2563			
		2066	50	103,275
106 + 50	1568			
		1111	50	55,563
107 + 00	655			
		328	50	16,375
107 + 50	0			
Total				1,374,826

1,374,826 ft^3 = 50,919 yd^3

6.5 ft × 10 ft, or 65 ft²,

$$\frac{49{,}500}{65} = 761.5$$

Say 770 holes. For the metric drill pattern 2 m × 3 m or 6 m²

$$\frac{4592.24}{6} = 765.4 \text{ holes}$$

Multiplying this figure by the average drilling depth plus 3.25 ft (1 m) of subdrilling we obtain the total linear feet of production drilling. We can determine the average depth of cut by totaling the cuts at all station cross sections in Figure 13.7. This average cut was given in Table 13.1.

Next we must calculate the total amount of presplit drilling required. To do this we refer to the specifications and see that the spacing between presplit holes is 3 ft (0.91 m) and any depth of cut over 6 ft (1.98 m) requires presplitting. Analyzing the cross section and site plan (Figure 13.6) we find that the linear distance requiring presplit is 940 ft (286.5 m), which requires 313 holes (by dividing the linear distance by the space between holes).

In Table 13.1 we see that the average depth of cut is 21.4 ft (6.53 m). Adding 3.25 ft (1 m) for subdrilling, we obtain the average borehole length. Thus:

$$21.4 \text{ ft} + 3.25 \text{ ft} = 24.65 \text{ ft}$$
$$(6.53 \text{ m} + 1 \text{ m} = 7.53 \text{ m})$$

Then:

$$24.65 \text{ ft} \times 770 = 18{,}980.5 \text{ ft of production drilling}$$
$$(7.53 \times 770 = 5798.1 \text{ m})$$

Average depth of cut is determined much in the same way as for production drilling. However, in the interest of brevity, we will assume an average depth of cut, including subdrilling, of 26 ft. However, since the presplit

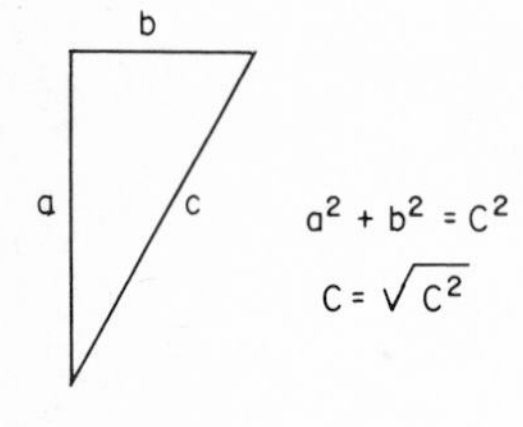

therefore:

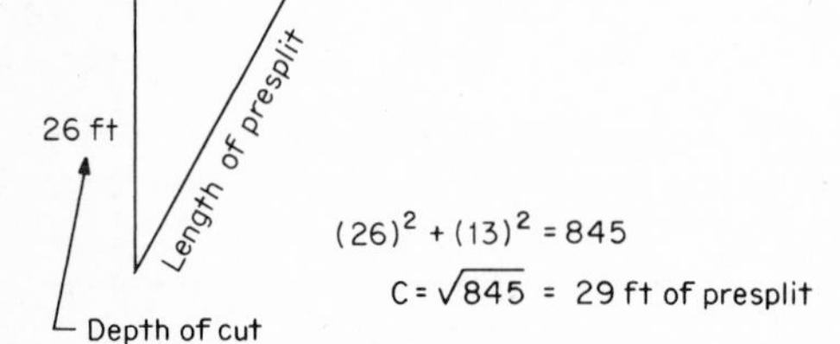

Fig. 13.8 Computing the length of the presplit hole.

holes are drilled on a slope, the actual length of the borehole is calculated by drawing a triangle and determining the hypotenuse. (See Figure 13.8.)

Now that we have 29 ft as the length of the presplit borehole, we multiply by the number of holes.

$$313 \times 29 \text{ ft} = 9077 \text{ (say 9100 ft)}$$
$$313 \times 8.83 \text{ m} = 2766.5 \text{ (say 2800 m)}$$

Now that all quantities have been determined the estimate can be constructed. We must decide as to the crew size and the equipment to be used. For the sake of this example, let us assume the use of four drills, two on production and two on presplit.

Crew Size

The equipment will require a crew of four drillers, one compressor operator, one blaster-foreman, and two laborers. Table 13.3 is an hourly labor cost breakdown.

$$\text{LABOR COST} = \$83.84/\text{h}$$

Equipment Costs

Next we have to determine the hourly equipment cost. Let us assume there are four drills at a total delivered price of \$225,000, with an estimated life of 5 yr at 2000 h/yr.

$$\text{HOURLY DEPRECIATION COST} = \frac{\$225{,}000}{10{,}000/\text{h}}$$
$$= \$22.50/\text{h}$$
$$\text{REPAIRS} = 50\% \text{ of depreciation}$$
$$= \$11.25/\text{h}$$
$$\text{AVERAGE ANNUAL INVESTMENT} = \frac{5 + 1}{2(5)} \times \$225{,}000$$
$$= \$135{,}000$$

For interest, insurance, and taxes and storage we'll assume 16%:

$$\frac{16\% \times \$135{,}000}{2000} = \$10.80/\text{h}$$

Assume that the lubrication costs are part of the compressor fuel, oil, and grease costs. For the compressors, assume that the air is banked and we require two 1200 ft^3/m machines at \$100,000 each. Total delivered price equals \$200,000.

$$\text{HOURLY DEPRECIATION} = \frac{\$200{,}000}{10{,}000 \text{ h}}$$
$$= \$20/\text{h}$$
$$\text{REPAIRS} = 50\% \times \$20 = \$10/\text{h}$$
$$\text{AVERAGE ANNUAL INVESTMENT} = \frac{(5 + 1) \times 200{,}000}{2(5)}$$
$$= \$120{,}000$$

TABLE 13.3 Labor Costs

Craft	Wage	Burden					Total
		FICA	Health and welfare	Insurance	Other	Number	
Blaster	$9.60	0.56	$1.30	.55	.15	1	$12.16
Operating engineer	9.50	0.55	1.50	.60	.15	1	12.30
Driller	7.80	0.46	1.30	.40	.15	4	40.44
Laborer	7.25	0.42	1.30	.35	.15	2	18.94
Total hourly labor cost							$83.84

For interest, taxes, insurance, and storage, we assume 16%:

$$\frac{16\% \times \$120{,}000}{2000} = \$9.60/\text{h}$$

For fuel, we assume 25 gallons per hour (gal/h), which is 94.6 liters per hour (L/h) at 50¢/gal (13.2¢/L).

$$\$0.50 \times 25 = \$12.50/\text{h}$$

For lubrication and maintenance, assume 30% of the fuel cost:

$$30\% \times 12.50 = \$3.75$$

When we total the hourly equipment costs, we find:

$$\text{TOTAL EQUIPMENT COST} = \$100.40/\text{h}$$

Totaling the hourly equipment and labor costs, we get:

$$\begin{aligned}\text{LABOR AND EQUIPMENT COST} &= \$100.40/\text{h} + 83.84/\text{h}\\ &= \$184.24/\text{h}\end{aligned}$$

Productivity

Next we can determine the time or production rate for the job. There are two different drilling activities: production drilling and presplit drilling. If we assume that the productivity rates for the two types of drilling are the same we can convert the total drilling into the time spent. If we assume that each drill can average 50 ft/h (15.2 m/h) then the total production for the four drills is 200 ft (60.9 m) per hour. Taking the total footage of presplit and production drilling and dividing it by the hourly production, we obtain the hours of drilling required.

18,980 ft of production drilling
+ 9,100 ft of presplit drilling

28,080 ft of drilling

$$\frac{28{,}080 \text{ ft}}{200 \text{ ft/h}} = 140.4 \text{ h (say 145 h)}$$

Metric:

5,798 m of production drilling
+ 2,800 m of presplit drilling

8,598 m of drilling

$$\frac{8{,}598 \text{ m}}{60.9 \text{ m/h}} = 141.2 \text{ h (say 145 h)}$$

To calculate the total equipment and labor costs one multiplies the hourly cost by the number of hours.

Hourly equipment cost	$100.40
Hourly labor cost	83.84
TOTAL HOURLY COST	$184.24

$$\$184.24/\text{h} \times 145 \text{ h} = \$26{,}714.80$$

The loading will be done during the drilling; therefore, the drill is the prime equipment (the drilling production is the true determinant).

Explosives Costs

Next we have to determine the material consumption and costs.

As discussed earlier in this chapter, the caps are determined by counting the number of blast holes and multiplying by a factor of 1.05. As previously determined, we have 770 blast holes, giving 808.5 caps. We'll say 810 blasting caps at a cost of 50¢ each.

$$\$0.50 \times 810 = \$405$$

It was determined that the presplit would be detonated by cord. Therefore, the total feet of presplit hole must be determined to find the detonating quantity. As calculated earlier there are 9100 ft (2800 m) of presplit hole. Therefore:

$$\begin{array}{rl} & 9{,}100 \text{ ft of presplit} \\ + & \underline{940} \text{ ft of trunk line} \\ & 10{,}040 \text{ ft of cord} \\ \times & \underline{1.20} \text{ (20\% waste)} \\ & 12{,}048 \text{ ft (say 12,000 ft)} \end{array}$$

At a price of 4.8¢/ft,

$$12{,}000 \times \$0.048 = \$576$$

Metric conversion:

$$\begin{array}{rl} & 2800 \text{ m of presplit} \\ + & \underline{286} \text{ m of trunk line} \\ & 3086 \text{ m of cord} \\ \times & \underline{1.20} \text{ (20\% waste)} \\ & 3703.2 \text{ m (say 3700 m)} \end{array}$$

$$3700 \text{ m} \times \$0.15/\text{m} = \$555$$

The next quantity to determine is the explosives required according to type.

The total of explosives required, using the volume found in Table 13.2, is:

$$1 \text{ lb/yd}^3 \times 50{,}919 \text{ yd}^3 = 50{,}919 \text{ lb}$$
$$589 \text{ g/m}^3 \times 38{,}933 \text{ m}^3 = 23{,}096 \text{ kg}$$

This quantity can be further broken down into types of explosives. In this author's experience, ammonium nitrate (AN/FO) is generally cheaper, and in many cases it will do as good a job as other explosives. Therefore, for this example, AN/FO will be utilized.

Assume that the holes are dry. Thus we can further assume that 25% of the explosives will be water gels and the remainder will be AN/FO.

Assuming dry holes is risky, so care must be taken not to overestimate the amount of AN/FO that may be substituted.

$$\begin{array}{r} 50{,}919 \text{ lb of explosives} \\ \times \quad 0.25 \\ \hline 12{,}730 \text{ lb (5774 kg) of water gels} \end{array}$$

leaving:

38,189 lb (17,322 kg)

PRESPLIT EXPLOSIVES = 0.45 lb/ft × 9100
= 4095 lb (1857 kg)

The explosives costs are:

Water gels: 12,730 lb × $0.55	$ 7,001.50
AN/FO: 38,189 lb × $0.12	4,582.68
Presplit: 4095 lb × $0.43	1,760.85
TOTAL	$13,345.03

Use 5% of the explosive costs to cover sundry items such as connecting wire, shooting box, and galvanometer. Bit and steel costs are generally per linear foot of drilling and are based on historical data. For this estimate use $0.30/ft ($0.98/m). Total drilling production and presplit times bit and steel cost yields:

28,080 ft × $0.30 = $8424
8598 m × $0.98 = $8426.04

Summarizing the costs thus far:

Equipment and labor	$26,714.80
Detonating cord	576.00
Blasting caps	405.00
Explosives	14,012.28
Bits and steel	8,424.00
TOTAL	$50,132.08

Since the job is too small to warrant an office trailer and other job overhead items, assume job overhead at 10%. Assume company overhead to equal 10% for profit margin. This yields:

Costs		$50,132.08
Job overhead, 10%	+	5,013.21
		$55,145.29
Company overhead, 10%	+	5,514.53
		$66,659.82
Profit, 10%	+	6,065.98
TOTAL BID PRICE:		$66,725.80

The unit bid price for this job is

$$\frac{\$66{,}725.80}{50{,}919\ \text{yd}^3} = \$1.31/\text{yd}^3$$

The costs and production rates in estimates will vary greatly from those put forth in this example. It is not the author's intent to set bidding parameters; quite the contrary. It is not recommended that any of the costs or production rates used in this example be used in an actual bid unless the cost or rate is first verified and determined to fit the particular bid situation.

STRATEGY

Once a contractor has determined the costs involved in a project, it is time to decide upon the allowance for profit. There are several things that have to be considered in making this decision, including the volume of work one currently has, the risk involved in doing the job, and the competition.

If all a contractor's crews are quite busy, it may be desirable to add a larger profit margin onto the job. In this way if the contractor does become overpriced and is not awarded the contract, there will be no harm, because there is still plenty of work; whereas if the job is awarded in spite of the larger profit margin, it will probably produce a very good profit or at least enough money that there will be no loss on the job. However, a contractor who needs work at the time to keep the workers employed may be willing to take a job at little or no profit to keep good help and support the overhead.

Another factor worth considering is the risk to which one will be exposed. If the job has a great deal of high-liability blasting, the contractor may either decline to bid on the job or put enough profit into the bid price to offset the risks involved. Also, if one is subcontracting, the contractor one would be working for may have a reputation for back-charging and being generally difficult to get along with. In this case a subcontractor may care to raise the profit margin to make the aggravation worth something. On the other hand, one may have an excellent working relationship with the contractor and therefore will be able to operate at a lower margin because there is less need for protection.

Finally, one has to consider what or who the competition is. If there is no competition on a job, there probably will be more latitude on profit margins. If one has a lot of competition, then the profit margin will have to be decreased. Also the nature of the competition can have an effect. If a job is large enough to attract large, out-of-town companies, then a local contractor will probably have lower mobilization costs and therefore may be able to increase the profit margin.

The factors affecting profit margin are all too numerous to adequately discuss in this text. However, the point that must be realized is that a company must have a bidding strategy for each proposal. Having a strategy is one way to help maximize profits and minimize risks.

In this chapter we discussed the steps in preparing an estimate, from quantity takeoff to bidding strategy. The main point to remember is that it is important for the engineer or contractor in addition to having a technical knowledge of drilling and blasting to also have a complete understanding of the considerations and costs involved in a blasting operation. Also, the methodology of developing an estimate is not important. What is important is that the person doing the estimate be aware of the work that has to be done and have a complete understanding of all costs involved in doing the project.

FOURTEEN

SAFETY

IMPORTANCE OF SAFETY

Not enough can be said about safety, particularly in blasting operations. Because of the dangers inherent in explosives use, the Institute of Makers of Explosives has established various recommended safe practices for blasting. These recommendations should be studied and followed.

RADIO-FREQUENCY ENERGY

The atmosphere around the antennas of radio transmitters and other radio-frequency (RF) energy producers, such as television and radar, contains fields of electrical energy created by the transmitters. These radio-frequency–produced electric fields can cause accidental detonation of electric blasting caps.

The intensity of the radio-frequency electric current in a blasting cap leg wire is dependent on the distance from the power source, the amount of re-acted power, the frequency, and the layout pattern of the blasting wire. The wires of a blasting layout can act as an antenna, depending on their configuration. If the wire patterns and the distance to the source antenna are compatible, the wire, acting as an antenna, can draw enough current from the air to detonate a cap. The current along the wire-pattern antenna will vary at different points, so the location of the cap in relation to the antenna has a bearing on whether it will accidentally detonate.

The type of transmitters most likely to accidentally detonate an electric blasting cap are commercial amplitude-modulated (AM) broadcast transmitters (0.535 to 1.605 megahertz, or MHz). The greater danger is the result of high power (aided by the relatively high altitudes of AM transmitters) and low frequency, sometimes combined with horizontal antennas (see Figure 14.1). The horizontal antenna, with its flat radiation pattern, creates more problems, because blasting circuits are horizontal. More current is induced at low frequencies than high frequencies. FM-frequency radio and television transmitters are generally not a problem for electric blasting. Mobile radios are hazardous only because they can be located right on the job. Because of

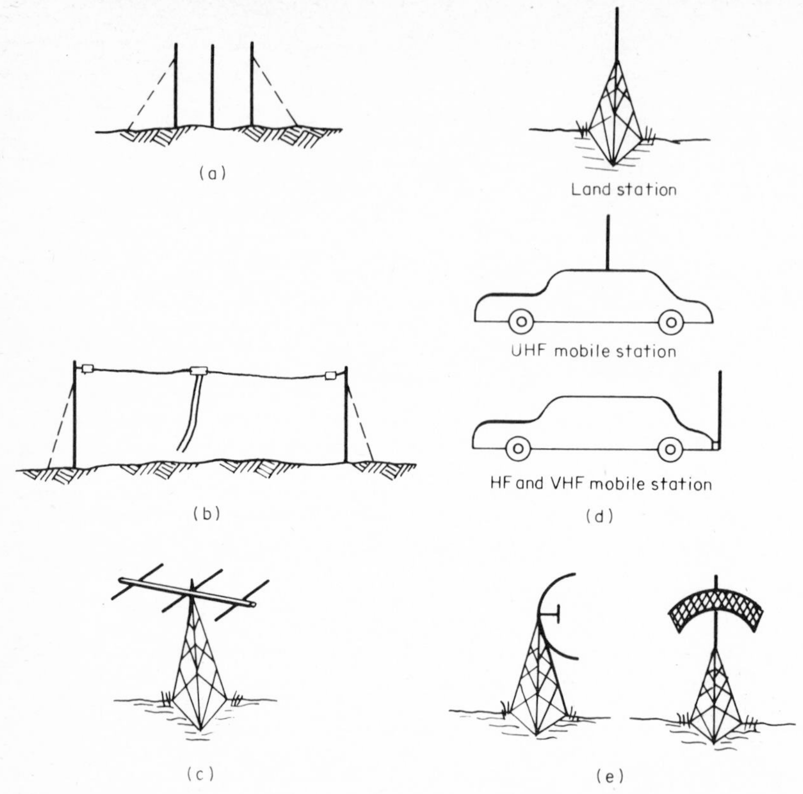

Fig. 14.1 Antenna types associated with the radio services: (*a*) AM broadcasting, vertically polarized; (*b*) high-frequency and international broadcasting, including some amateur, horizontally polarized; (*c*) amateur high-frequency beam antenna; (*d*) mobile service; (*e*) radar service. (*Institute of Makers of Explosives.*)

the very high frequency, radiation patterns, and locations, the problem created by microwave relays and other microwave devices is minimal. However, because of the RF uncertainties, underground operations, such as mines, should be very carefully structured. Current heats the bridge wire regardless of frequency, and 50 milliamperes (mA) will generate the same heat whether it is dc, 60 hz, TV, or microwave.

Greatest Hazards

The greatest hazards exist from AM transmitters, under the following conditions:

1. If there is a straight length of wire equal to one-half the radio wavelength or a multiple thereof, with the cap in the center (wavelength in feet equals 1000 divided by the frequency in megahertz)

2. If there is a straight length of wire equal to one-quarter of the radio wavelength or an odd multiple thereof (i.e., times 3,5,7, etc., times the wavelength)
3. If there is a straight length of wire that is perpendicular to the bearing (straight line) from the job to antenna
4. If the wire is off the ground by a few feet

Figure 14.2 illustrates the first two conditions.

Before a blasting project begins the site should be tested for RF energy. If the sources of potential RF energy are known, Tables 14.1 to 14.6 can be used, together with Figures 14.3 to 14.6. However, if the sources are unknown or there are several sources, using the tables is difficult, and a test will have to be made.

Tests

An ammeter or pilot test bulb can be used to test for RF energy. An RF ammeter will respond to frequencies up to 100 MHz. The pilot test bulb is a good field test because it will respond to all frequencies and simulates an electric-cap bridge wire.

A pilot test bulb, such as General Electric No. 48, will glow white at 60 mA and will have a dim orange glow at 40 mA. The IME recommends 50 mA as the safe limit for electric blasting. The half-wave dipole is the best-suited test pickup for FM and television. It is the most accurate test circuit, consist-

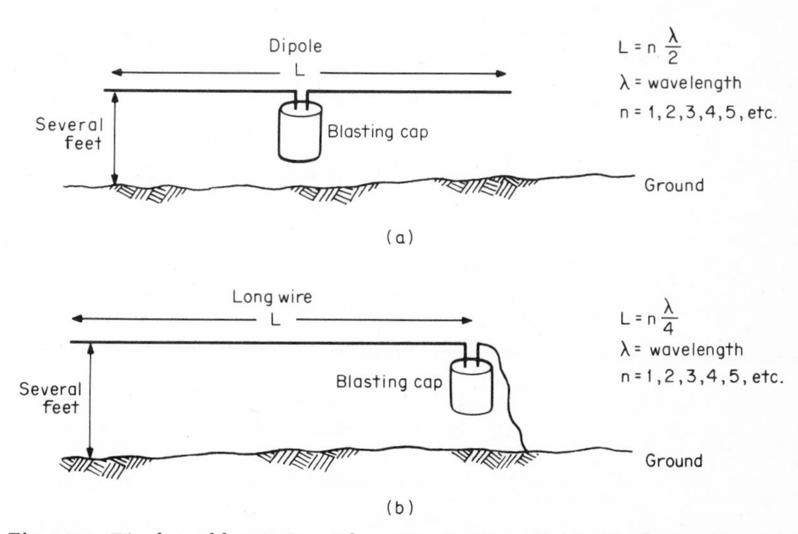

Fig. 14.2 Dipole and long-wire pickup circuits. Two pilot test bulb circuits used to simulate receiving antennae: (*a*) one-half wave dipole and (*b*) one wire grounded to earth, the other grounded several feet above the ground. (*Institute of Makers of Explosives.*)

TABLE 14.1 Recommended Distances from Commercial AM Broadcast Transmitters, 0.535 to 1.605 MHz (Fig. 14.3)

Transmitter power,* W	Minimum distance, ft
Up to 4,000	750
5,000	850
10,000	1,200
25,000	2,000
50,000†	2,800
100,000	3,900
500,000	8,800

* Power delivered to antenna.

† 50,000 W is the present maximum power of U.S. broadcast transmitters in this frequency range.

SOURCE: "Safety Guide for the Prevention of Radio Frequency Radiation Hazards in the Use of Electric Blasting Caps," Institute of Makers of Explosives, New York, 1978, p. 10.

ing of a straight piece of wire one-half the wavelength (or any multiple thereof), with the bulb and the socket at the center of the wire. If the wavelength is unknown, use a dipole of 32 ft to simulate a blasting cap with 16-ft leg wires. With the wire stretched out at 90° to the direction of the antennas, watch the bulb for glow (see Figure 14.7); then cut 1 to 2 ft from each end of

TABLE 14.2 Recommended Distance from Transmitters up to 30 MHz (Excluding AM Broadcast) Calculated for a Specific Loop Pickup Configuration*† (Fig. 14.4)

Transmitter power,† W	Minimum distance, ft
100	750
500	1,700
1,000	2,400
5,000	5,500
50,000	17,000
500,000‡	55,000

* This table should be applied to International Broadcast Transmitters in the 10–25 MHz range.

† Power delivered to antenna.

‡ Present maximum for International Broadcast.

SOURCE: "Safety Guide for the Prevention of Radio Frequency Radiation Hazards in the Use of Electric Blasting Caps," Institute of Makers of Explosives, New York, 1978, p. 10.

TABLE 14.3 Recommended Distances from Mobile Transmitters Including Amateur and Citizens' Band

Transmitter[a] power, W	MF 1.6 to 3.4 MHz Industrial	HF 28 to 29.7 MHz Amateur	VHF 35–36 MHz Public Use 42–44 MHz Public Use 50–54 MHz Amateur	VHF 144 to 148 MHz Amateur 150.8–161.6 MHz Public Use	UHF 450–470 MHz Public Use
10	40	100	40	15	10
50	90	220	90	35	20
100	125	310	130	50	30
180[b]				65	40
250	200	490	205	75	45
500[c]			290		
600[d]	300	760	315	115	70
1,000[e]	400	980	410	150	90
10,000[f]	1,250		1,300		

Citizens Band, Class D Transmitters, 26.96–27.41 MHz

Type	Recommended minimum distance: Hand-held	Recommended minimum distance: Vehicle-mounted
Double-sideband, 4 W maximum transmitter power	5 ft	15 ft
Single-sideband, 12 W peak envelope power	20 ft	60 ft

[a] Power delivered to antenna.

[b] Maximum power for two-way mobile units in VHF (150.8 or 161.6 MHz range) and for two-way mobile and fixed-station units in UHF (450 to 460 MHz range).

[c] Maximum power for major VHF two-way mobile and fixed station units in 35 to 44 MHz range.

[d] Maximum power for two-way fixed station units in VHF (150.8 to 161.6 MHz range).

[e] Maximum power for amateur radio mobile units.

[f] Maximum power for some base stations in 42 to 44 MHz band and 1.6 to 1.8 MHz band.

SOURCE: "Safety Guide for the Prevention of Radio Frequency Radiation Hazards in the Use of Electric Blasting Caps," Institute of Makers of Explosives, New York, 1978, p. 11.

TABLE 14.4 Recommended Distances from VHF TV and FM Broadcasting Transmitters (Fig. 14.5)

Effective radiated power, W	Minimum distance, ft	
	Channels 2 to 6 and FM	Channels 7 to 13
Up to 1,000	1,000	750
10,000	1,800	1,300
100,000*	3,200	2,300
316,000†	4,300	3,000
1,000,000	5,800	4,000
10,000,000	10,200	7,400

* Present maximum power channels 2 to 6 and FM: 100,000 W.
† Present maximum power channels 7 to 13: 316,000 W.
SOURCE: "Safety Guide for the Prevention of Radio Frequency Radiation Hazards in the Use of Electric Blasting Caps," Institute of Makers of Explosives, New York, 1978, p. 12.

the wire and observe the bulb. This should be repeated until the wire has about 2 ft remaining. For testing for AM, one side of the dipole should be grounded and the other wire should be several feet above the ground and extend several hundred feet.

When you use caps with leg wires made of iron and longer than 8 ft, half the safe distance given in Tables 14.1 to 14.6 is required. Iron leg wires

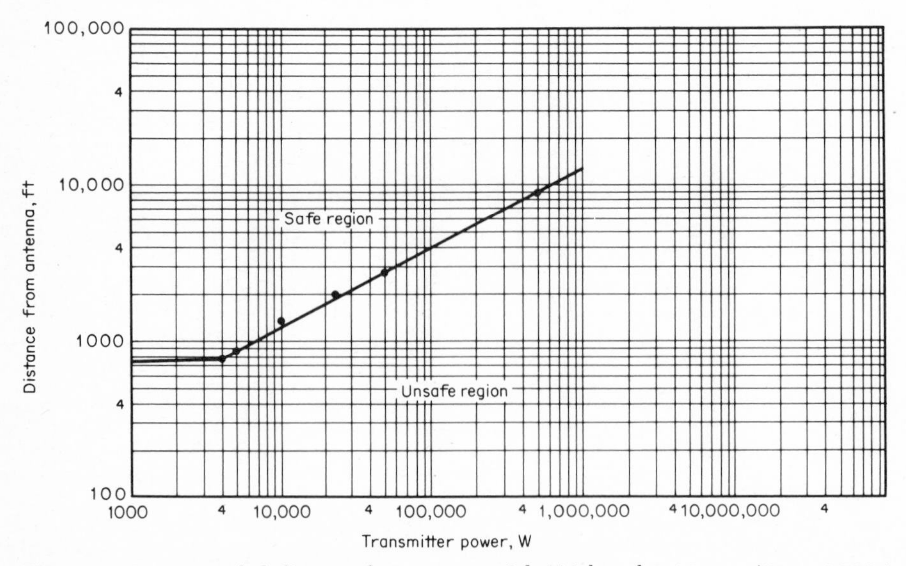

Fig. 14.3 Recommended distance from commercial AM broadcast transmitters—0.535 to 1.605 MHz. (*Institute of Makers of Explosives.*)

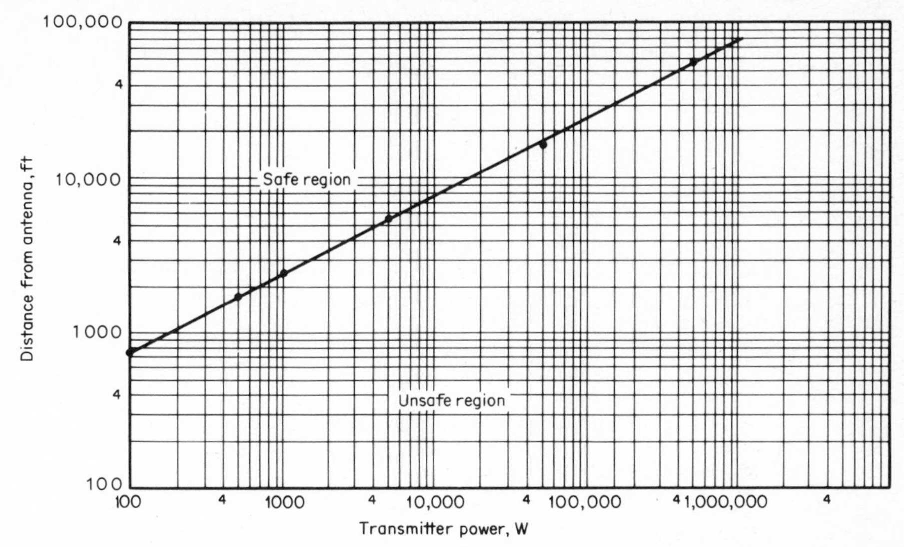

Fig. 14.4 Recommended distance from transmitters up to 30 MHz. (*Institute of Makers of Explosives.*)

have 6 times the resistance to RF energy of copper leg wires. If the blasting circuit contains both copper and iron wires, use the iron wire near the cap bridge wire to take advantage of its higher resistance.

STRAY GROUND CURRENTS

When blasting with electric blasting caps one must watch for stray ground currents. If lightning strikes it will detonate part or all of the shot. Stray

TABLE 14.5 Recommended Distances from UHF TV Transmitters (Fig. 14.6)

Effective radiated power, W	Minimum distance, ft
Up to 10,000	600
1,000,000	2,000
5,000,000*	3,000
100,000,000	6,000

* Present maximum power channels 14 to 83: 5 MW.

SOURCE: "Safety Guide for the Prevention of Radio Frequency Radiation Hazards in the Use of Electric Blasting Caps," Institute of Makers of Explosives, New York, 1978, p. 12.

TABLE 14.6 Recommended Distances from Maritime Radionavigational Radar

Service	Effective radiated power, W	Minimum distance, ft
Small pleasure craft	500 (3 cm)	20
Harbor craft, river boats, etc.	5,000 (3 cm)	50
Large commercial shipping	50,000 (3 cm and 10 cm)	300

The above table should be used only if the exact nature of the radar hazard is understood. In cases where an uncertainty exists as to the nature of the radar signal as well as ground scatter and reflection of the radar signal, a recommended minimum distance of 1000 ft should be maintained from the radar antenna.

SOURCE: "Safety Guide for the Prevention of Radio Frequency Radiation Hazards in the Use of Electric Blasting Caps," Institute of Makers of Explosives, New York, 1978, p. 17.

ground currents are more likely near conductors, such as streams, railroad tracks, pipelines, and power lines. One of the most expedient methods for monitoring lightning activity in the field is to have an AM radio set between stations; electrical storms can be heard in the form of increased static. If a lightning storm approaches remove everyone from the area. If the project is in an area subject to many electrical storms it is a good idea to load the holes in a sequence that can be cut off and shot quickly: begin loading at the face,

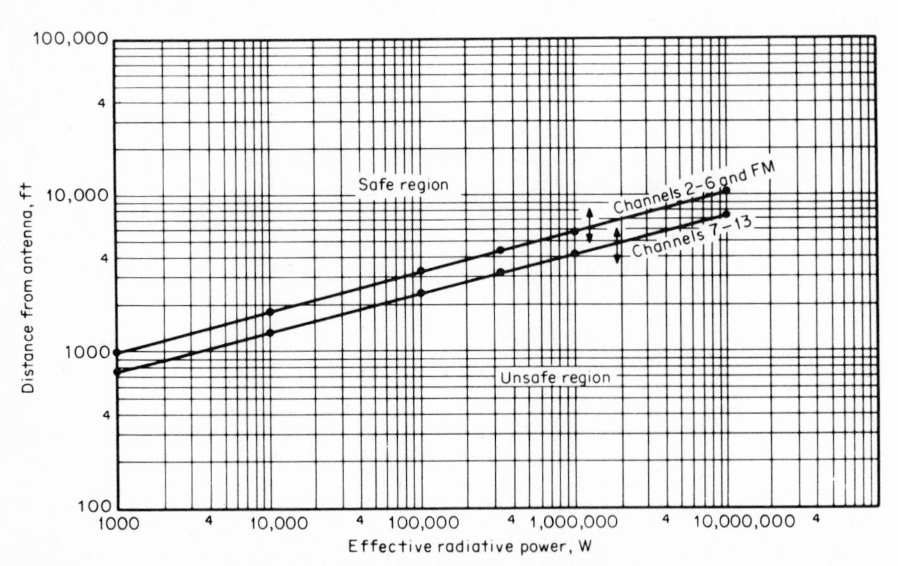

Fig. 14.5 Recommended distance from VHF TV and FM transmitters. (*Institute of Makers of Explosives.*)

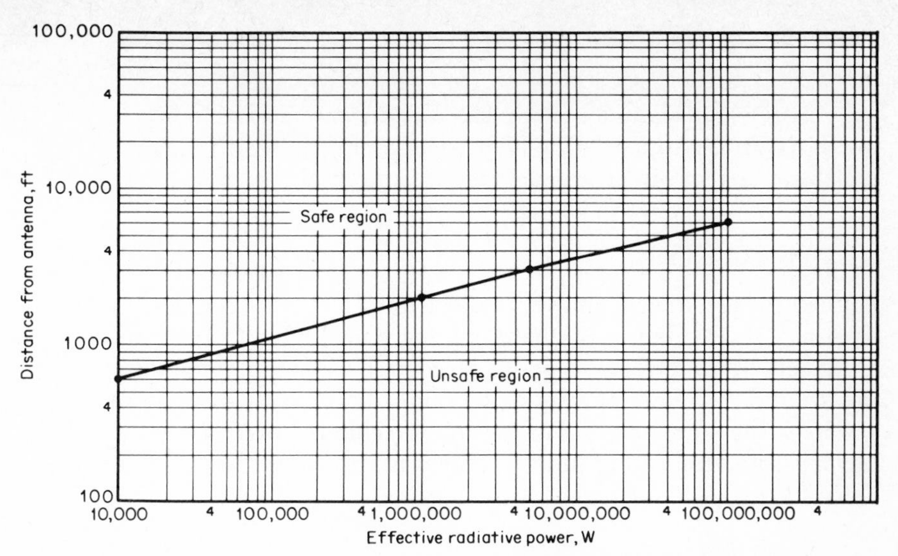

Fig. 14.6 Recommended distance from UHF TV transmitters. (*Institute of Makers of Explosives.*)

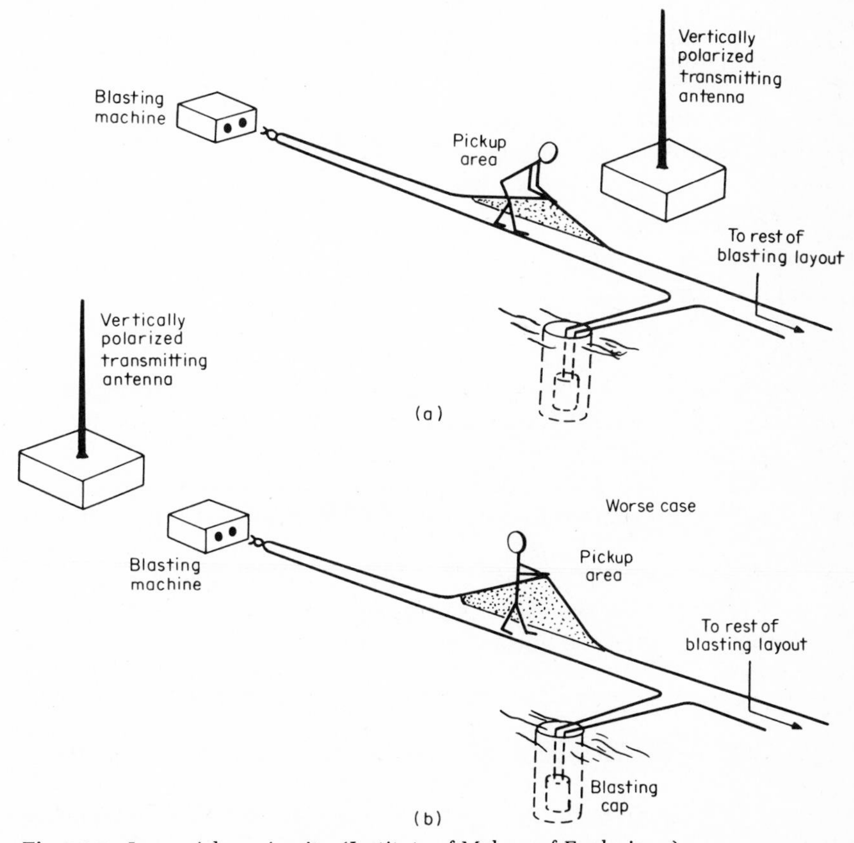

Fig. 14.7 Loop pickup circuits. (*Institute of Makers of Explosives.*)

keeping the loading in a block pattern and stemmed immediately after loading. If a storm is approaching it may be possible to fire the partial shot before the storm arrives, avoiding considerable danger.

STATIC ELECTRICITY

Premature initiation from static electricity is more common with seismic blasting. This is due to the long cap leg wires, which can store more static electricity. Even a worker's body can act as a capacitor, depending on insulation from the ground, body size, and how dry the skin is. If these conditions are favorable for acting as a capacitor, the charge intensifies until the worker touches something to ground it. If that is an electric blasting cap, it can be detonated.

Conditions Conducive to Static

Conditions conducive to static electricity are found in the vicinity of dust storms or snowstorms, at high altitudes, at low humidity, and near mechanical equipment such as moving conveyor belts. To avoid mechanically induced static all equipment and equipment parts should be connected at a common point and grounded through a resistance of less than 1 Ω. All connections should be taped, and electric blasting caps or wires from the blasting circuit should be kept clear of the machinery. Also keep the machinery grounding wire away from railroad tracks, pipes, or any other source of stray current.

Pneumatic Conveyance The pneumatic conveyance of solids (such as AN/FO) is a source of static electricity. This can be controlled by using a conductive system, a system which will dissipate the static electricity instead of building it up. The loaders should be grounded by direct contact with a wet surface, by being attached to a metal stake in the ground. When underground the system may be attached to a rock bolt or by putting a scaling bar, a metal bar used for sealing loose rock from the face, into an empty borehole. Keep the grounding away from track or pipelines because of the possibility of sources of stray current. Use semi-conductive hoses and be sure the tank that contains the AN/FO is also grounded. The effect that you are trying to create is a free path through which the static charges that become built up on the AN/FO particles can be dissipated to the ground through the walls of the borehole. Therefore, the use of plastic hole liners is forbidden.

STRAY CURRENTS

The IME has established that the levels of stray currents should be below 50 mA, which is one-fifth the amperage required to detonate a cap.

Current from batteries, generators, or power lines to any electrical equipment always tries to return to its source through paths or conductors. These sources of possible stray current, as mentioned earlier, include railroad tracks, pipelines, and streams. Before using electric blasting caps, measure the stray currents in the area. Except for wet earth and highly conductive strata separated by narrow nonconductive strata, the earth is not a good conductor of electricity.

Testing

To test for stray currents an ac/dc voltmeter capable of reading 0.05 V should be used. Use one specifically designed for blasting, because the type used by electricians can detonate a cap. One end of the testing wire should be connected to a metal grounding stake driven into moist earth (pour water there to make it moist, if necessary). Using a wire with no more resistance than the shortest cap leg wire, attach the free end to the object being tested. Test everything in the blast area, moving the grounding stake to each location for best results. Be sure to test for both alternating current and direct current. The same procedure is used for both, but only one can be done at a time.

GALVANIC ACTION

Dissimilar metals in contact with each other, either directly or through a conductive medium, act as a battery. That is, mixing metals creates voltage. These are low voltages but are still capable of detonating an electric cap. This was realized in seismic prospecting when blasting started using aluminum loading poles. The aluminum poles, which were contained in steel casings, in alkaline drilling produced enough current to detonate a cap. Therefore, only wooden or nonmetallic loading poles should be used. Also, in salt water blasting particular attention has to be given to galvanic action, because salt water is a conductor.

"DOS AND DON'TS"

The "dos and don'ts" are a listing of suggested rules for the safe handling and use of explosives adopted by the Institute of Makers of Explosives February 1, 1964. They are reproduced here, and an explanation or clarification of some of the rules will be given after each grouping.

Definitions The term "explosives" as used herein includes any or all of the following: dynamite, black blasting powder, pellet powder, blasting caps, electric blasting caps, and detonating cord.

The term "electric blasting cap," as used herein, includes both instantaneous electric blasting caps and all types of delay electric blasting caps. The term "primer," as used herein, means a cartridge of explosives in combination with a blasting cap or an electric blasting cap.

When Transporting Explosives DO obey all federal, state and local laws and regulations. DO see that any vehicle used to transport explosives is in proper working condition and equipped with tight wooden or non-sparking metal floors with sides and ends high enough to prevent the explosives from falling out. The load in an open-bodied truck should be covered with a waterproof and fire-resistant tarpaulin, and the explosives should not be allowed to contact any source of heat such as an exhaust pipe. Wiring should be fully insulated so as to prevent short circuiting, and at least two fire extinguishers should be carried. The trucks should be plainly marked so as to give adequate warning to the public of the nature of the cargo.

DON'T permit metal, except approved metal truck bodies, to contact cases of explosives. Metal, flammable, or corrosive substances should not be transported with explosives.

DON'T allow smoking or unauthorized or unnecessary persons in the vehicle.

DO load and unload explosives carefully. Never throw explosives from the truck.

DO see that other explosives, including detonating cord, are separated from blasting caps and/or electric blasting caps where it is permitted to transport them in the same vehicle.

DON'T drive trucks containing explosives through cities, towns, or villages, or park them near such places as restaurants, garages, and filling stations, unless it cannot be avoided.

DO request that explosive deliveries be made at the magazine or in some other location well removed from populated areas.

DON'T fight fires after they have come in contact with explosives. Remove all personnel to a safe location and guard the area against intruders.

The foregoing incorporate the Department of Transportation (DOT) regulations and general rules of safety concerning the transportation of explosives. The intent is to avoid fire and shock, either of which could cause a detonation of the load.

When Storing Explosives DO store explosives in accordance with federal, state or local laws and regulations.

DO store explosives only in a magazine which is clean, dry, well ventilated, reasonably cool, properly located, substantially constructed, bullet and fire resistant, and securely locked.

DON'T store blasting caps or electric blasting caps in the same box, container or magazine with other explosives.

DON'T store explosives, fuse, or fuse lighters in a wet or damp place, or near oil, gasoline, cleaning solution, or solvents, or near radiators, steam pipes, exhaust pipes, stoves, or other sources of heat.

DON'T store any sparking metal, or sparking metal tools in an explosive magazine.

DON'T smoke or have matches, or any source of fire or flame in or near an explosives magazine.

DON'T allow leaves, grass, brush, or debris to accumulate within 25 ft of an explosives magazine.

DON'T shoot into explosives or allow the discharge of firearms in the vicinity of an explosives magazine.

DO consult the manufacturer if nitroglycerin from deteriorated explosives has leaked onto the floor of a magazine. The floor should be desensitized by washing thoroughly with an agent approved for that purpose.

DO locate explosives magazines in the most isolated places available. They should be separated from each other, and from inhabited buildings, highways, and railroads, by distances not less than those recommended in the "American Table of Distances."

The storing of explosives is regulated by various agencies; however, they all use the *American Table of Distances* as their guide (see Tables 14.7 and 14.8). Again it is obvious that the rules are to avoid contact of the explosives and fire and shock (concussion).

When Using Explosives DON'T use sparking metal tools to open kegs or wooden cases of explosives. Metallic slitters may be used for opening fiberboard cases, provided that the metallic slitter does not come in contact with the metallic fasteners of the case.

DON'T smoke or have matches or any source of fire or flame within 100 ft of an area in which explosives are being handled or used.

DON'T place explosives where they may be exposed to flame, excessive heat, sparks, or impact.

Appendices DO replace or close the cover of explosives cases or packages after using.

DON'T carry explosives in the pockets of your clothing or elsewhere on your person.

DON'T insert anything but fuse in the open end of a blasting cap.

DON'T strike, tamper with, or attempt to remove or investigate the contents of a blasting cap or an electric blasting cap, or try to pull the wires out of an electric blasting cap.

DON'T allow children or unauthorized or unnecessary persons to be present where explosives are being handled or used.

DON'T handle, use, or be near explosives during the approach or progress of any electrical storm. All persons should retire to a place of safety.

DON'T use explosives or accessory equipment that is obviously deteriorated or damaged.

DON'T attempt to reclaim or to use fuse, blasting caps, electric blasting caps, or any explosives that have been water soaked, even if they have dried out. Consult the manufacturer.

These recommendations involving using explosives are quite self-explanatory. It should be noted that the blasting caps must be handled with even greater care than explosives.

When Preparing the Primer DON'T make up primers in a magazine, or near excessive quantities of explosives, or in excess of immediate needs.

TABLE 14.7 American Table of Distances for Storage of Explosives (as revised and approved by the Institute of Makers of Explosives, November 5, 1971)

Quantity of explosives		Distances, ft: Inhabited buildings		Public highways Class A to D		Passenger railways; public highways with traffic volume of more than 3000 vehicles/day		Separation of magazines	
Pounds over	Pounds not over	Barricaded	Unbarricaded	Barricaded	Unbarricaded	Barricaded	Unbarricaded	Barricaded	Unbarricaded
2	5	70	140	30	60	51	102	6	12
5	10	90	180	35	70	64	128	8	16
10	20	110	220	45	90	81	162	10	20
20	30	125	250	50	100	93	186	11	22
30	40	140	280	55	110	103	206	12	24
40	50	150	300	60	120	110	220	14	28
50	75	170	340	70	140	127	254	15	30
75	100	190	380	75	150	139	278	16	32
100	125	200	400	80	160	150	300	18	36
125	150	215	430	85	170	159	318	19	38
150	200	235	470	95	190	175	350	21	42
200	250	255	510	105	210	189	378	23	46
250	300	270	540	110	220	201	402	24	48
300	400	295	590	120	240	221	442	27	54
400	500	320	640	130	260	238	476	29	58
500	600	340	680	135	270	253	506	31	62
600	700	355	710	145	290	266	532	32	64
700	800	375	750	150	300	278	556	33	66
800	900	390	780	155	310	289	578	35	70
900	1,000	400	800	160	320	300	600	36	72
1,000	1,200	425	850	165	330	318	636	39	78
1,200	1,400	450	900	170	340	336	672	41	82
1,400	1,600	470	940	175	350	351	702	43	86
1,600	1,800	490	980	180	360	366	732	44	88
1,800	2,000	505	1,010	185	370	378	756	45	90
2,000	2,500	545	1,090	190	380	408	816	49	98
2,500	3,000	580	1,160	195	390	432	864	52	104
3,000	4,000	635	1,270	210	420	474	948	58	116
4,000	5,000	685	1,370	225	450	513	1,026	61	122
5,000	6,000	730	1,460	235	470	546	1,092	65	130

TABLE 14.7 American Table of Distances for Storage of Explosives (as revised and approved by the Institute of Makers of Explosives, November 5, 1971) (*Continued*)

Quantity of explosives		Distances, ft							
		Inhabited buildings		Public highways Class A to D		Passenger railways; public highways with traffic volume of more than 3000 vehicles/day		Separation of magazines	
Pounds over	Pounds not over	Barri- caded	Un- barri- caded	Barri- caded	Un- barri- caded	Barri- caded	Un- barri- caded	Barri- caded	Un- barri- caded
6,000	7,000	770	1,540	245	490	573	1,146	68	136
7,000	8,000	800	1,600	250	500	600	1,200	72	144
8,000	9,000	835	1,670	255	510	624	1,248	75	150
9,000	10,000	865	1,730	260	520	645	1,290	78	156
10,000	12,000	875	1,750	270	540	687	1,374	82	164
12,000	14,000	885	1,770	275	550	723	1,446	87	174
14,000	16,000	900	1,800	280	560	756	1,512	90	180
16,000	18,000	940	1,880	285	570	786	1,572	94	188
18,000	20,000	975	1,950	290	580	813	1,626	98	196
20,000	25,000	1,055	2,000	315	630	876	1,752	105	210
25,000	30,000	1,130	2,000	340	680	933	1,866	112	224
30,000	35,000	1,205	2,000	360	720	981	1,962	119	238
35,000	40,000	1,275	2,000	380	760	1,026	2,000	124	248
40,000	45,000	1,340	2,000	400	800	1,068	2,000	129	258
45,000	50,000	1,400	2,000	420	840	1,104	2,000	135	270
50,000	55,000	1,460	2,000	440	880	1,140	2,000	140	280
55,000	60,000	1,515	2,000	455	910	1,173	2,000	145	290
60,000	65,000	1,565	2,000	470	940	1,206	2,000	150	300
65,000	70,000	1,610	2,000	485	970	1,236	2,000	155	310
70,000	75,000	1,655	2,000	500	1,000	1,263	2,000	160	320
75,000	80,000	1,695	2,000	510	1,020	1,293	2,000	165	330
80,000	85,000	1,730	2,000	520	1,040	1,317	2,000	170	340
85,000	90,000	1,760	2,000	530	1,060	1,344	2,000	175	350
90,000	95,000	1,790	2,000	540	1,080	1,368	2,000	180	360
95,000	100,000	1,815	2,000	545	1,090	1,392	2,000	185	370
100,000	110,000	1,835	2,000	550	1,100	1,437	2,000	195	390
110,000	120,000	1,855	2,000	555	1,110	1,479	2,000	205	410
120,000	130,000	1,875	2,000	560	1,120	1,521	2,000	215	430
130,000	140,000	1,890	2,000	565	1,130	1,557	2,000	225	450
140,000	150,000	1,900	2,000	570	1,140	1,593	2,000	235	470

TABLE 14.7 American Table of Distances for Storage of Explosives (as revised and approved by the Institute of Makers of Explosives, November 5, 1971) (*Continued*)

Quantity of explosives		Distances, ft							
		Inhabited buildings		Public highways Class A to D		Passenger railways; public highways with traffic volume of more than 3000 vehicles/day		Separation of magazines	
Pounds over	Pounds not over	Barri-caded	Un-barri-caded	Barri-caded	Un-barri-caded	Barri-caded	Un-barri-caded	Barri-caded	Un-barri-caded
150,000	160,000	1,935	2,000	580	1,160	1,629	2,000	245	490
160,000	170,000	1,965	2,000	590	1,180	1,662	2,000	255	510
170,000	180,000	1,990	2,000	600	1,200	1,695	2,000	265	530
180,000	190,000	2,010	2,010	605	1,210	1,725	2,000	275	550
190,000	200,000	2,030	2,030	610	1,220	1,755	2,000	285	570
200,000	210,000	2,055	2,055	620	1,240	1,782	2,000	295	590
210,000	230,000	2,100	2,100	635	1,270	1,836	2,000	315	630
230,000	250,000	2,155	2,155	650	1,300	1,890	2,000	335	670
250,000	275,000	2,215	2,215	670	1,340	1,950	2,000	360	720
275,000	300,000	2,275	2,275	690	1,380	2,000	2,000	385	770

SOURCE: "The American Table of Distances," Institute of Makers of Explosives, New York, 1977, pp. 6–7.

DON'T force a blasting cap or an electric blasting cap into dynamite. Insert the cap into a hole made in the dynamite with a punch suitable for the purpose.

DO make up primers in accordance with proven and established methods. Make sure that the cap shell is completely encased in the dynamite or booster and so secured that in loading no tension will be placed on the wires or fuse at the point of entry into the cap. When side priming a heavy wall or heavy weight cartridge, wrap adhesive tape around the hole punched in the cartridge so that the cap cannot come out.

Primers are quite hazardous to handle because the sensitivity of the blasting cap is present with the explosive force of the explosive.

When Drilling and Loading DO comply with applicable federal, state and local regulations relative to drilling and loading.

DO carefully examine the surface or face before drilling to determine the possible presence of unfired explosives. Never drill into explosives.

TABLE 14.8 Table of Recommended Separation Distances of Ammonium Nitrate and Blasting Agents from Explosives or Blasting Agents

Donor weight		Minimum separation distance of acceptor when barricaded, ft		Minimum thickness of artificial barricades, in
Pounds over	Pounds not over	Ammonium nitrate	Blasting agent	
	100	3	11	12
100	300	4	14	12
300	600	5	18	12
600	1,000	6	22	12
1,000	1,600	7	25	12
1,600	2,000	8	29	12
2,000	3,000	9	32	15
3,000	4,000	10	36	15
4,000	6,000	11	40	15
6,000	8,000	12	43	20
8,000	10,000	13	47	20
10,000	12,000	14	50	20
12,000	16,000	15	54	25
16,000	20,000	16	58	25
20,000	25,000	18	65	25
25,000	30,000	19	68	30
30,000	35,000	20	72	30
35,000	40,000	21	76	30
40,000	45,000	22	79	35
45,000	50,000	23	83	35
50,000	55,000	24	86	35
55,000	60,000	25	90	35
60,000	70,000	26	94	40
70,000	80,000	28	101	40
80,000	90,000	30	108	40
90,000	100,000	32	115	40
100,000	120,000	34	122	50
120,000	140,000	37	133	50
140,000	160,000	40	144	50
160,000	180,000	44	158	50
180,000	200,000	48	173	50
200,000	220,000	52	187	60
220,000	250,000	56	202	60
250,000	275,000	60	216	60
275,000	300,000	64	230	60

SOURCE: NFPA 492, Boston, 1976. By permission from National Fire Protection Association, Boston.

DO check the borehole carefully with a wooden tamping pole or measuring tape to determine its condition before loading.

DO recognize the possibility of static electrical hazards from pneumatic loading and take adequate precautionary measures. If any doubt exists, consult your explosives supplier.

DON'T stack surplus explosives near working areas during loading.

DO cut from the spool the line of detonating cord extending into a borehole before loading the remainder of the charge.

DON'T load a borehole with explosives after springing (enlarging the hole with explosives) or upon completion of drilling without making certain that it is cool and that it does not contain any hot metal or burning or smoldering material. Temperatures in excess of 150°F are dangerous.

DON'T spring a borehole near another hole loaded with explosives.

DON'T force explosives into a borehole or through an obstruction in a borehole. Any such practice is particularly hazardous in dry holes and when the charge is primed.

DON'T slit, drop, deform or abuse the primer. DON'T drop a large size, heavy cartridge directly on the primer.

DO avoid placing any unnecessary part of the body over the borehole during loading.

DON'T load any borehole near electric power lines unless the firing line, including the electric blasting cap wires, is so short that it cannot reach the power wires.

DON'T connect blasting caps or electric blasting caps to detonating cord except by methods recommended by the manufacturer.

As with all the dos and don'ts these tips on drilling and loading are advice on how to prevent accidental detonation of the explosives.

When Tamping DON'T tamp dynamite that has been removed from the cartridge.

DON'T tamp with metallic devices of any kind, including the metal end of loading poles. Use wooden tamping tools with no exposed metal parts, except non-sparking metal connectors for jointed poles. Avoid violent tamping. Never tamp with the primer.

DO confine the explosives in the borehole with sand, earth, clay, or other suitable incombustible stemming material.

DON'T kink or injure fuse or electric blasting cap wires, when tamping.

You must be even more careful when tamping the primer charge; you are better off not to tamp it at all.

When Shooting Electrically DON'T uncoil the wires or use electric blasting caps during dust storms or go near any other sources of large charges of static electricity.

DON'T uncoil the wires or use electric blasting caps in the vicinity of radio-frequency transmitters, except at safe distances. Consult the manufacturer or the Institute of Makers of Explosives pamphlet "Radio Frequency Hazards."

DO keep the firing circuit completely insulated from the ground or other conductors such as bare wires, rails, pipes, or other paths of stray currents.

DON'T have electric wires or cables of any kind near electric blasting caps or other explosives except at the time and for the purpose of firing the blast.

DO test all electric blasting caps, either singularly or when connected in a series circuit, using only a blasting galvanometer specifically designed for the purpose.

DON'T use in the same circuit electric blasting caps made by more than one manufacturer or electric blasting caps of different style or function even if made by the same manufacturer unless such use is approved by the manufacturer.

DON'T attempt to fire a single electric blasting cap or a circuit of electric blasting caps with less than the minimum current specified by the manufacturer.

DO be sure that all wire ends to be connected are bright and clean.

DO keep the electric cap wires or leading wires disconnected from the power source and short circuited until ready to fire.

Study the section on radio-frequency hazards earlier in this chapter and these other precautions.

When Shooting with Fuse DO handle fuse carefully to avoid damaging the covering. In cold weather warm slightly before using to avoid cracking the waterproofing.

DON'T use short fuse. Know the burning speed of the fuse and make sure you have time to reach a place of safety after lighting. Never use less than 2 ft.

DON'T cut fuse until you are ready to insert it into a blasting cap. Cut off an inch or two to insure a dry end. Cut fuse squarely across with a clean sharp blade. Seat the fuse lightly against the cap charge and avoid twisting after it is in place.

DON'T crimp blasting caps by any means except a cap crimper designed for the purpose. Make certain that the cap is securely crimped to the fuse.

DO light fuse with a fuse lighter designed for the purpose. If a match is used the fuse should be slit at the end and the match head held in the slit against the powder core. Then scratch the match head with an abrasive surface to light the fuse.

DON'T light fuse until sufficient stemming has been placed over the explosive to prevent sparks or flying match heads from coming into contact with the explosive.

DON'T hold explosives in the hands when lighting fuse.

In Underground Work DO use permissible explosives only in the manner specified by the United States Bureau of Mines.

DON'T take excessive quantities of explosives into a mine at any one time.

DON'T use black blasting powder or pellet powder with permissible explosives or other dynamite in the same borehole in a coal mine.

Before and after Firing DON'T fire a blast without a positive signal from the one in charge, who has made certain that all surplus explosives are in a safe place, all persons and vehicles are at a safe distance or under sufficient cover, and that adequate warning has been given.

DON'T return to the area of any blast until the smoke and fumes from the blast have been dissipated.

DON'T attempt to investigate a misfire too soon. Follow recognized rules and regulations, or if no rules or regulations are in effect, wait at least one hour.

DON'T drill, bore, or pick out a charge of explosives that has misfired. Misfires should be handled only by, or under the direction of, a competent and experienced person.

Explosives Disposal DON'T abandon any explosives.

DO dispose of or destroy explosives in strict accordance with approved methods. Consult the manufacturer or follow the Institute of Makers of Explosives pamphlet on destroying explosives.

DON'T leave explosives, empty cartridges, boxes, liners, or other materials used in the packing of explosives lying around where children or unauthorized persons or livestock can get at them.

DON'T allow any wood, paper, or any other materials employed in packing explosives to be burned in a stove, a fireplace, or other confined places, or to be used for any purpose. Such materials should be destroyed by burning at an isolated location out-of-doors and no person should be nearer than 100 ft after the burning has started.

The dos and don'ts contain a considerable amount of good advice in a particularly condensed list. It is important to study and understand them. There is no excuse for not having the knowledge, because a copy of the dos and don'ts generally comes in all cases of explosives and boxes of blasting caps.

FIFTEEN

SPECIFICATIONS AND INSPECTION

SPECIFICATIONS

Specifications are the written instructions that supplement the drawings. They define the quality and workmanship required on the job and are generally given greater legal strength: if there is a discrepancy between the drawings and the specifications, the specifications will generally override the drawings.

Classification

The clauses in the specifications are classified as general provisions and technical provisions. The general provisions apply to the work as a whole. Descriptions of the work, materials to be furnished by others, and references to the contract drawings are topics. The technical provisions describe the technical details of each type of construction and should contain the detailed instructions necessary to obtain the desired quality of workmanship. There should be one section to cover each type of construction. The drilling and blasting specifications may be found in the earthwork section, in the excavation section, or, if the blasting is a major portion of the contract, in a section devoted to drilling and blasting.

If the project has both surface and tunnel, or underground construction, the specifications will be divided into tunnel and surface. However, these independent specifications may refer to one another.

Writing Specifications

In writing the technical specifications, the concern should be with the finished product. Under certain conditions the methods may be dictated; however, it is best to state the required results and leave the method to the contractor. In the blasting specification the finished requirement may be the maximum size of breakage, the line and grade, and public safety.

The primary problems with stating methods, particularly with blasting, are twofold: (1) the restrictions may be unwarranted; and (2) the net result may be the inflating of costs. The only time methods can be stipulated is when the specification writer is very knowledgeable in the field. This occurrence is extremely rare where blasting is concerned. However, the specification can give limitations and restrictions concerning the methods for the sake of coordination. That is, the specification may require more than one move-in for the blasting contractor because of schedule and coordinating problems in preparing areas for blasting. If this is foreseen it should be explained or required in the specifications.

Also the specification should give limitations on the effects of the blasting on adjacent property or structures. The various laws generally give such limitations, but the specifications should refer to these laws or even contain the actual instructions. These restrictions may concern the amount of tolerable vibration, through either actual measurement or calculation based on the scaled distance formula; or they may stipulate the requirement for blasting mats.

With regard to blasting mats, or other covering, there should be certain parameters established, but the requirement should be referred back to the engineer, who can then decide whether mats are unnecessary on a given shot and can permit the blast without mats. The reason for the alternative for matting is cost: the more matting required, the more costly the blasting. This does not mean that the engineer should be careless, but to have a blanket requirement for blasting mats is costly and demonstrates a definite lack of knowledge of the subject.

In establishing the parameters for presplitting, the results should be the primary concern. Therefore, to establish a required presplit, spacing without regard to changes in rock type is counterproductive. The goals of the project are best served if a test shot of the required presplit design is first required, with the engineer being permitted to approve changes in the presplit design. The advantages to this are numerous. First, the contractor who is doing the blasting is ultimately responsible for the quality of the presplit, whereas if the procedure is dictated then the one who dictates the method has to accept the responsibility for the project. Second, the more competent blaster is rewarded for expertise; that is, a blaster who is quite knowledgeable about presplitting will know from the test shot whether the conditions will make it permissible to spread out the spacing. That blaster has a competitive edge in being able to lower the bid because of the lower presplit costs. Thus rewarding the competent blaster also lowers the cost of the project. It is important that the specification presplit requirements be strict so that the responsibility for the presplitting results lies with the contractor.

The specification writer should establish minimum acceptable standards or qualifications for the person who will supervise the loading of explosives. These qualifications may include a blasting license, in locations that require or issue blasting licenses, and possibly a résumé of experience and references to be submitted to the engineer for approval.

Example of a Specification

The following is an example of a blasting specification. The purpose of this example is to demonstrate the type of restrictions and requirements that a blasting specification may contain. This specification should not be taken literally, because it may be too restrictive for some projects and not restrictive enough for others. However, it should give the reader a feel for a blasting specification.

201 Blasting

A. General

1. Blasting shall be done to permit the cut to the lines, grades, and cross sections indicated on the drawings.
2. Rock shall be fragmented to less than one cubic yard. Breakage one cubic yard or greater will be further fragmented by mechanical means or blockholing at no expense to the owner.

B. Blasting Competence

1. All blasting will be done by competent experienced blasters.
2. The contractor will submit a résumé of experience and references of the person or persons responsible for the supervision of blasting for the engineer's approval.
3. Persons responsible for blasting shall be present and supervise all blast design, loading, and shot firing.
4. Persons responsible for blasting shall have the required license and conduct all blasting operations in accordance with applicable laws and safety regulations.

C. Amounts of Explosives

1. Delay blasting methods will be used.
2. Explosives will be used in weights not to exceed a particle velocity of one inch per second when measured or not to exceed the amount that can be detonated within any 8-ms period obtained by the following formula:

$$W = (D/60)^2$$

where W = weight of explosive, lb
D = distance, ft

where the distance is the linear distance to the nearest structure or project perimeter, whichever is lower.

3. The engineer may require that blasts be monitored with a seismograph at the sole expense of the contractor.

D. Mats

1. Within 300 ft of any structure on the project boundaries, mats will be used.
2. At distances of greater than 300 ft to structures or project boundaries the engineer may require mats at no additional expense to the engineer or owner.
3. Backfilling will not be allowed unless approved by the engineer.

E. Presplitting

1. Presplitting will be done at those lines indicated on the drawings that have a depth of cut of 6 ft or greater from the finished elevation to the top of the cut.
2. All presplit holes will be drilled true to the required angle.
3. All presplit holes will be 3 in in diameter.
4. For the first shot involving presplit holes the spacing between the holes will be 30 in and continually loaded with ⅞-in diameter explosives cartridges to within 3 ft of the top of the nearest free face. If the results of the test blast indicate overbreak, then the spacing may be extended at the approval of the engineer.

When writing a specification for presplitting another alternative is to leave the design completely to the discretion of the contractor, but to hold the contractor responsible for the line. For example, the specification can be such that the contractor has a certain tolerance, that if the rock on the presplit line intervenes into the excavation over a certain limit, say 6 in, the contractor has to chip the "tights"; and that if the overbreak is over a certain limit, the contractor, in the case of a foundation, has to replace the rock with concrete. This method is quite strict and should be used only when the presplit tolerances are quite important.

INSPECTION

When assigned the task of inspection on a blasting project, the engineer, as with other facets of the job, is primarily interested in adherence to the plans and specifications. All matters concerning the interpretation of the specification should be tempered with the knowledge of the project goals. That is, as the job progresses, if the specification ends up being unreasonable or incorrect, seek to have it changed.

In addition to watching for adherence to plans, specifications, and safety regulations, the engineer may be required to check the credentials of the blaster. Checking the blaster's license is a convenient way to do this; and even if the area in which the project is located does not require blasters to be licensed, they may still be licensed elsewhere. Another source for checking blasters' competence is through their insurance carriers, who will have opinions on this matter based on loss records. Also, the regulatory agency—the state's safety office or bureau of mines—that has jurisdiction over the project area may have information or an opinion of the blasters. If a résumé and references are required, the references may be checked.

The writing of specifications and the inspection of blasting projects is at best difficult for engineers who lack much blasting experience. However, if both of these tasks are done in light of the project goals and public safety, then the interest of the owner has been served.

SIXTEEN

REGULATORY AGENCIES

HISTORY

Every facet of blasting is regulated by one government agency or another. Federal, state, municipal, and county governments regulate explosives from production to consumption.

Until the end of the 1960s the only federal government agencies involved in the blasting industry were the Interstate Commerce Commission and the Bureau of Mines. However, since then more federal government agencies have become involved in the explosives industry, in addition to increased regulations at local levels. (See Table 16.1.)

Before the growth in regulation the Interstate Commerce Commission (ICC) regulated the packaging and transportation of explosives and the Bureau of Mines regulated the storage and some of the safety practices in the use of explosives. The two regulatory agencies came from two different departments, the Departments of Transportation and the Interior, respectively.

Since that time the Departments of Labor and the Treasury have become involved. The Department of Labor involvement in the blasting industry is in the form of the Williams-Steiges Occupational Safety and Health Act of 1970. This act established the Occupational Safety and Health Administration of the Department of Labor (OSHA), which has the power to mandate rules and regulations regarding working conditions and the power to impose strict penalties on those deemed in violation of the regulations. The act affects the use of explosive material in both general industry and construction.

DEPARTMENT OF LABOR

The Department of Labor also added the Mine Safety and Health Administration (formerly the Department of the Interior's Mine Enforcement Safety Administration, MESA). This agency is responsible for health and safety in mines, both underground and surface. Their regulations cover the use of

TABLE 16.1 Federal Control of Explosive Materials

Agency	Responsibility	Code of Federal Regulations	Regulations cover:
Department of Transportation, Office of Hazardous Material Operations (DOT)	Safe transportation of hazardous materials via highway, rail, air, and water	49 CFR, Parts 107* and 171 through 178	Explosive material definitions, classification, packaging, marking and labeling, and transportation requirements for rail, aircraft, vessel, and public highway
Department of the Treasury, Bureau of Alcohol, Tobacco and Firearms (ATF)	Storage of and commerce in explosive materials to prevent their illegal use	27 CFR, Part 181	Requirements for storage and record keeping and restrictions on purchase, sales, and distribution of explosive materials
Department of Labor, Mining Safety and Health Administration (MSHA)	Health and safety of underground and surface miners	30 CFR, Parts 15 and 16	Criteria for permissible explosives and stemming devices
		24 and 25	Criteria for permissible blasting machines
		55	Use of explosive materials in metal and nonmetal surface mines

		56	Use of explosive materials in sand, gravel, and stone operations
		57	Use of explosive materials in metal and nonmetal underground mines
		75	Use of explosive materials in underground coal mines
		77	Use of explosive materials in surface coal mines and surface work areas of underground coal mines
Department of Labor, Occupational Safety and Health Administration (OSHA)	Health and safety of construction and general industry workers	29 CFR, Part 1910	Use of explosive materials in general industry
		1926	Use of explosive materials in construction
Department of Interior, Office of Surface Mining Reclamation and Enforcement (OSM)	To protect general public and the environment from the effects of coal mining	30 CFR, Part 715	The environmental effects of blasting, such as ground vibration, air blast, flyrock, etc., and the safety of the general public in the surrounding area

* There are additional DOT regulations covering the operation of transportation equipment in 49 CFR that also apply to the movement of explosive materials.

explosive materials in metal and nonmetal surface mines, underground coal mines, sand, gravel, and surface coal mines, and surface work areas of underground coal mines. Also, the agency establishes the criteria for permissible blasting machines and permissible explosives and stemming devices for use in coal mines.

DEPARTMENT OF THE INTERIOR

The Department of the Interior, Office of Surface Mining Reclamation and Enforcement (OSM) is trusted with the responsibility of protecting the general public and the environment from the effects of coal mining. This includes regulating the effects on the environment from blasting. OSM regulations restrict the amount of ground vibration, air blast, and flyrock, and they define the obligations of the persons responsible for blasting for the safety of the general public in the area of the blast.

DEPARTMENT OF TRANSPORTATION

The Department of Transportation is responsible for the transportation of hazardous materials, whether the conveyance is rail, water, air, or highway. It regulates packaging, material definitions and classifications, and the requirements for marking and labeling.

DEPARTMENT OF THE TREASURY

The Department of the Treasury's Bureau of Alcohol, Tobacco and Firearms (ATF) entered the regulation of explosives via the Organized Arms Control Act of 1970. ATF is primarily interested in the illegal use of explosives. However, the blasting industry is regulated by the ATF in the areas of storage and commerce of explosive materials in the effort to prevent illegal use, primarily to keep explosives out of the hands of terrorists. The ATF regulates the blasting industry by establishing restrictions on the purchase, sale, and distribution of explosives and the requirements for storage and record keeping.

OVERLAPPING

One might think that there would be overlapping of the regulations imposed by these various federal agencies. Well, that is the case. Although they all have different purposes they tend to overlap on different subject matters. For example, the Department of Transportation is concerned with the transportation of explosive materials. However, the following excerpts are from the OSHA Rules and Regulations (*Federal Register,* Vol. 39, No. 122). You will note that the regulation cites the Department of Transportation Regulations governing the transportation of explosive materials and then proceeds with OSHA's restrictions.

§1926.902 Surface transportation of explosives.

a Transportation of explosives shall meet the provisions of Department of Transportation regulations contained in 14 CFR Part 103, Air Transportation; 46 CFR Parts 146–149, Water Carriers; 49 CFR Parts 171–179, Highways and Railways; 49 CFR Part 180, Pipelines; and 49 CFR Parts 390–397, Motor Carriers.

b Motor vehicles or conveyances transporting explosives shall only be driven by, and be in the charge of, a licensed driver who is physically fit. He shall be familiar with the local, state, and federal regulations governing the transportation of explosives.

c Vehicles used for transporting explosives shall be strong enough to carry the load without difficulty, and shall be in good mechanical condition.

The maintaining of inventory records is required by both the ATF and OSHA. ATF:

§181.127 Daily summary of magazine transactions.

In taking the inventory required by §181.122, §181.123, §181.124, and §181.125, the inventory shall be entered in a record of daily transactions to be maintained at each magazine of an approved storage facility. At the close of business each day each licensee and permittee shall record by class of explosive materials, as prescribed in the Explosive List, the total quantity received in and removed from each magazine during the day and the total remaining on hand at the end of the day. Any discrepancy which might indicate a theft or loss of explosive materials shall be reported in accordance with the provisions of 181.130.

OSHA:

All explosives shall be accounted for at all times. Explosives not being used shall be kept in a locked magazine, unavailable to persons not authorized to handle them. The employer shall maintain an inventory and use record of all explosives. Appropriate authorities shall be notified of any loss, theft, or unauthorized entry into a magazine.

Finally, the various agencies have their requirements for blaster qualifications. OSM:

c All blasting operations shall be conducted by experienced, trained, and competent persons who understand the hazards involved. Each person responsible for blasting operations shall—

1. Have demonstrated a knowledge of, and shall comply with, MSHA safety requirements and U.S. Department of Treasury security requirements;
2. Be capable of using mature judgment in all situations;
3. Be in good physical condition and not addicted to intoxicants, narcotics, or other similar types of drugs;

4. Possess current knowledge of the local, state, and federal laws and regulations applicable to the work; and
5. Possess a valid certificate of completion of training and qualifications as required by 30 CFR 850 and 851.

OSHA:

§1926.901 Blaster qualifications.

a A blaster shall be able to understand and give written and oral orders;

b A blaster shall be in good physical condition and not be addicted to narcotics, intoxicants, or similar types of drugs;

c A blaster shall be qualified, by reason of training, knowledge, or experience, in the field of transporting, storing, handling, and use of explosives, and have a working knowledge of state and local laws and regulations which pertain to explosives.

d Blasters shall be required to furnish satisfactory evidence of competency in handling explosives and performing in a safe manner the type of blasting that will be required.

e The blaster shall be knowledgeable and competent in the use of each type of blasting method used.

The agencies also overlap on storage requirements. However, these requirements are based on the recommendations of the Institute of Makers of Explosives.

It should be obvious that it is important for all persons engaged in blasting activities to be aware of the laws that govern them.

To gain a further appreciation of the amount and the sources of regulation, we shall follow explosives from manufacturing to consumption, showing the effects of the various federal regulations (see Figure 16.1). The agencies that regulate the manufacture of the explosives are ATF, DOT, and OSHA.

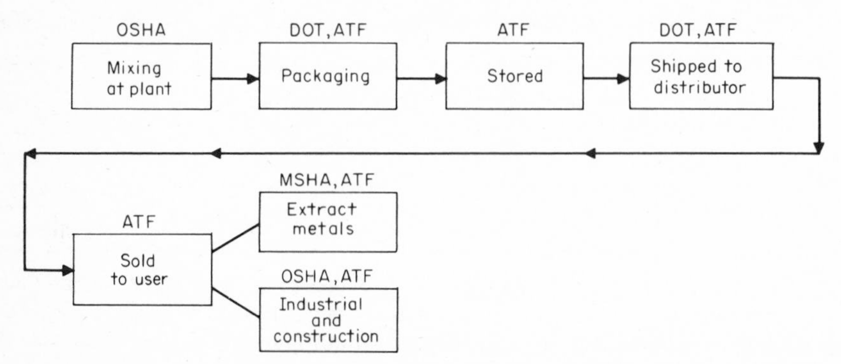

Fig. 16.1 Flowchart of federal regulations governing explosives from manufacturing to sale to an interstate user.

As the raw materials enter the explosive plant, OSHA rules and regulations oversee the safety of the workers. After the materials are processed and form an explosive compound, the DOT regulates the construction of the cartridge shells, bags, and boxes. DOT regulations also state how the packages are to be marked concerning the contents. Then the ATF requires each cartridge and each container to have a code marking that identifies the manufacturer, the plant, the shift, and the date it was made. This code is the inventory number for all recording use—including the records of the user. Also, the ATF requires that a record of the explosive be maintained by the manufacturer from the time the product is marked until it is delivered to the user. The user must maintain a record of the explosive from the time delivery is taken until the time the explosive is consumed, or detonated.

After packaging, the explosives are either stored according to ATF regulations or shipped according to DOT regulations to a distributor or user, either of which has to store or continue to ship according to the ATF or DOT, respectively. The sale of explosives from the distributor to the user comes under the regulation of the ATF, and the delivery is regulated by the DOT. The local user, the one who deals only intrastate, is exempt from ATF licensing requirements and record keeping. However, the storage is still regulated by local regulatory agencies. These agencies, like the federal agencies, tend to follow the recommendation of the Institute of Makers of Explosives.

LOCAL REGULATIONS

Most states have their own laws governing the use of explosive materials. The regulations generally come under the department of safety (as in Massachusetts) or labor and are enforced by these agencies. In states that have heavy mining activity the regulation of blasting and explosives may come under the department of mines (for example, Kentucky). The type of regulations will vary from state to state, with a range of very little regulation to quite heavy regulation.

Probably the greatest variation in these regulations of blasting activities is within the state between municipalities. For example, two towns that are adjacent to each other can differ tremendously in blasting restrictions. Generally the more severely regulated town is showing an overreaction to a bad experience with a blasting operation.

One major problem with blasting regulations at the municipal level is that regulations are written by well-meaning people who have absolutely no idea of proper blasting procedures. There are some towns that put such restrictions on blasting that for a small project (e.g., a ledge in a swimming pool) the cost can be increased by 500%. This author has obtained blasting permits in towns that had blasting restrictions that were dangerous. In one case, the town required a blaster to secure the blasting mat so that it would not move when the shot was fired. What this means is that a blaster had to chain the corners of the mat to steel stakes in the ground to prevent the mat from moving. Of course it doesn't take much thought to realize that the result was that a missile was constructed that would be catapulted through the air when the inertia pulled the stake out of the ground.

Most state and local regulations follow federal law; that is, many local regulatory agencies will adapt the federal regulations. When blasting in an area, the general rule is: obey the regulations that are the strictest. Therefore, if there is a difference in the allowable particle velocity, follow the most conservative.

LICENSING OF BLASTERS

The licensing of blasters is just beginning to become a federal requirement in some cases. Until recently all licenses were local requirements. Generally the state will require and administer the licensing of blasters (as do Massachusetts, California, and Kentucky). However, in some states licenses are not required by the state but municipalities may require a license. Arizona is an example of this: Arizona does not require that a blaster be licensed, but the City of Phoenix does require it. Therefore, Phoenix has its own licensing procedure.

CONCLUSIONS

No one likes to be regulated, whether it be an individual or industry. However, most would agree that some kind of regulation on explosives is needed to protect the public at large. Whether or not the current level of regulation of explosives is too much, too little, or unrealistic is subject for debate. However, the best way for an industry to avoid regulation is to upgrade and educate its members. The problem with restrictions is that the competent are often restricted along with the incompetent.

APPENDIXES

TABLE A.1 Explosives Products (Atlas Powder Company)

Product	Typical density,* g/cm³	Product	Typical density,* g/cm³
NG Explosives		Aquanal Q	1.22
Power Primer	1.36	Aquanal W	1.23
60% Giant Gelatin	1.43	Powermax 100	1.20†
40% Giant Gelatin	1.53	Powermax 200	1.20†
Gelodyn #1	1.29	Powermax 355	1.25
Gelodyn #3	1.22	Apex 463	1.22
Gelodyn #4	1.09	Apex 700	1.22
Gelodyn #5	1.03	AN/FO	
Powerdyn	1.35	Pellite	0.81
Ammodyte	1.20	Pellite (pneumatic loaded)	0.98
60% Extra Dynamite	1.29	Pellite HD	1.10
40% Extra Dynamite	1.35	Pellite HDAL	1.10
Prestex A	1.18		Approx. lb.
Slurries		Presplitting	exp./foot
Aquaflo	1.25	Kleen-Kut C	0.45
Aquaram	1.20	Kleen-Kut D	0.32
Aquanal	1.22	Kleen-Kut E	0.25

* Density may vary slightly dependent on style and diameter.
† Style PE.
Trademarks of Atlas Powder Company include Ammodyte, Gelodyn, Power Primer, Prestex, Aquanal, Aquaram, Pellite, Aquaflo, Kinepak, and Aquagel.

TABLE A.2 Rockmaster® 38 SF Millisecond Delays (Atlas Powder Company)

Delay no.	Average time, ms	Delay no.	Average time, ms
1	8	20	1000
2	25	21	1125
3	50	22	1250
4	75	23	1375
5	100	24	1500
6	125	25	1625
7	150	26	1750
8	175	27	1875
9	200	28	2000
10	250	29	2125
11	300	30	2250
12	350		
13	400		
14	450		
15	500		
16	550		
17	650		
18	750		
19	875		

All Atlas EB caps shown here, including INSTANTANEOUS, can be intermixed in the same circuit because the bridge wire resistances are the same.

TABLE A.3 Timemaster® SF ½-s Delays (Atlas Powder Company)

Delay no.	Average time, ms
0	8
1	500
2	1000
3	1500
4	2000
5	2500
6	3000
7	3500
8	4000
9	4500
10	5000
11	5500
12	6000
13	6500
14	7000
15	7500

TABLE A.4 Kolmaster® SF Delays (Atlas Powder Company)

Delay no.	Average time, ms	Delay no.	Average time, ms
1	25	6	350
2	100	7	400
3	175	8	450
4	250	9	500
5	300	10	550

"Safety First" (SF) design provides substantial protection from static and arcing. Atlas was the first U.S. manufacturer to add this extra safety feature to delay caps.

TABLE A.5 Resistance of Atlas Copper Wire Electric Blasting Caps (Atlas Powder Company)

Length of leg wire, ft	Nominal resistance, Ω	Length of leg wire, ft	Nominal resistance, Ω
8	1.7	40	2.3
10	1.7	50	2.6
12	1.8	60	2.8
16	1.9	80	3.3
20	2.1	100	3.8
24	2.3	120	4.4
30	2.5	150	5.1

TABLE A.6 Nominal Resistance of Copper Leading Wire (Atlas Powder Company)

AWG	Ω/1000 ft	AWG	Ω/1000 ft
No. 4	0.248	No. 14	2.52
No. 6	0.395	No. 16	4.02
No. 8	0.628	No. 18	6.38
No. 10	0.999	No. 20	10.15
No. 12	1.59	No. 22	16.14

TABLE A.7 Squares, Cubes, Square Roots, and Cube Roots

n	n^2	n^3	$\sqrt{n}$	$\sqrt{10n}$	$\sqrt[3]{n}$	$\sqrt[3]{10n}$	$\sqrt[3]{100n}$
1	1	1	1.000 000	3.162 278	1.000 000	2.154 435	4.641 589
2	4	8	1.414 214	4.472 136	1.259 921	2.714 418	5.848 035
3	9	27	1.732 051	5.477 226	1.442 250	3.107 233	6.694 330
4	16	64	2.000 000	6.324 555	1.587 401	3.419 952	7.368 063
5	25	125	2.236 068	7.071 068	1.709 976	3.684 031	7.937 005
6	36	216	2.449 490	7.745 967	1.817 121	3.914 868	8.434 327
7	49	343	2.645 751	8.366 600	1.912 931	4.121 285	8.879 040
8	64	512	2.828 427	8.944 272	2.000 000	4.308 869	9.283 178
9	81	729	3.000 000	9.486 833	2.080 084	4.481 405	9.654 894

GLOSSARY OF BLASTING TERMS

AN/FO A blasting agent that is a mixture of ammonium nitrate and fuel oil.

Back The term given to the roof of an underground excavation, such as a tunnel or cavern.

Back Break The breakage that occurs beyond the last row of blast holes.

Bench The horizontal plane on which vertical blast holes are drilled.

Benching Excavation of material in levels or stepped terraces.

Binary Explosive A two-component explosive in which the two components are not considered explosives until mixed (generally at the site).

Blast The act of fragmenting rock by the use of explosives.

Blasting Agent Any material or mixture that consists of fuel and oxidizer which is not classified as an explosive and none of whose ingredients is classified as an explosive. When packaged for shipment, the finished product cannot be detonated by a No. 8 blasting cap.

Blast Hole A hole drilled for the placement of explosives in rock or other material to be blasted.

Blockholing A method of fragmenting boulders by drilling a hole into their mass and placing explosives into the hole.

Booster An explosive used to increase the intensity of another explosive.

Bootleg The portion of the blast hole that remains relatively in place after the blast; generally due to insufficient explosives for the amount of burden to be moved.

Borehole See blast hole.

Bottom-Priming A method in which the primer is placed at the bottom of the column of explosives.

Bridging The interruption of the continuity of the explosive column.

Brisance The ability of an explosive to break or shatter material by shock. This characteristic is more common in military applications.

Bulk Strength The strength of a cartridge of explosives compared to the same-sized cartridge of straight-nitroglycerin dynamite.

Burden The distance from the explosive to the nearest free face.

Collar The blast hole opening.

Collar-Priming Placing the primer at the top of the explosive column.

Connecting Wire Wire used to connect to and extend the length of leg wire or lead wire.

Connector A device used on detonating cord to initiate a delay.

Coyote Blasting A method of driving tunnels into the rock face at the toe and loading the tunnels with explosives.

Cushion Blasting A method of blasting that will leave a neat excavation line by firing a single row of holes to shear the web between closely drilled holes.

Cut The depth and width dimensions of a face.

Decking Separating a portion of the blast hole from the main charge by stemming, air gap, or void.

Delay Blasting The use of devices to cause a delay in the detonation of explosive charges.

Delay Blasting Cap A blasting cap with a built-in delay device that allows various charges to be detonated at different times.

Detonating Cord A plastic cord with a core of high-velocity explosives used to detonate explosive charges.

Fragmentation The extent to which the rock is broken or reduced in size.

Galvanic Action The generation of electric currents caused when dissimilar metals are in contact with each other or are in a conductive medium (e.g., salt water).

Galvanometer A gauge to measure resistance in an electric blasting circuit.

Initiate The act of detonating high explosives by a mechanical device or other means.

Joints Planes of weakness in rock masses that offer no resistance to separation.

Jumbo A drilling rig designed to have two or more drills mounted on it. Common for shafts and tunnels.

Lead Wire (Shooting Line) Wire used to connect an electric blasting circuit to the power source.

Leg Wires Wires extending from an electric blasting cap.

Lifters The bottom holes of a tunnel round.

Mat Matting or netting used to contain flyrock. Usually made of wire rope or rubber tires.

Metallization The adding of metal powders, usually aluminum or ferrosilicon, to a blasting agent to sensitize or booster.

Misfiring Failure of a portion or all of a shot to detonate when initiated. Can be very dangerous!

Muck Broken or fragmented rock resulting from a blast.

Mudcapping A technique for blasting boulders whereby the charge is placed on the surface of the boulder and covered with mud.

Overbreak Rock fragmented or broken beyond the desired excavation line.

Overburden Material, usually dirt or gravel, that lies on top of the rock to be blasted.

Oxidizer The provider of oxygen, such as ammonium nitrate in AN/FO.

Permissible Explosives permitted for use in underground coal mines by the U.S. Bureau of Mines.

Powder Slang term for all solid explosives.

Presplitting Detonation of a single row of holes on the excavation line before the production blast to leave a neat excavation line by shearing the web between the holes.

Primary or Production Blast Primary blast necessary to maintain production or the blast that is the payload.

Primer An explosive used to initiate an entire explosive charge.

Pull The amount of advance in a heading by drilling and blasting.

Round A pattern of drill holes that amounts to a unit of advance in underground blasting.

Sequential Firing A method of firing holes in a sequence to reduce the burden and provide many independent blasts.

Shunt Connection of two cap leg wires together to prevent accidental detonation from stray currents.

Spacing The distance between boreholes in a row.

Stemming Material placed in the blast hole from the top of the explosive column to the collar to prevent gases from escaping upon detonation.

Strength The explosive energy of an explosive relative to a like amount of nitroglycerin dynamite.

Subdrilling Drilling holes below the required elevation to ensure breakage to the required elevation.

Tamping The act of compressing the explosive in the blast hole.

TNT Trinitrotoluene, the explosive to which all military explosives are compared.

Toe The distance between the bottom of the blast hole and the free face.

Velocity The rate, or speed, at which the detonation wave travels through an explosive.

INDEX